HOW MAPS CHANGE THINGS

Ward L. Kaiser

HOW MAPS CHANGE THINGS

A conversation about
the maps we choose
and the world we want

CopperHouse

Cover design: Simon Loffler, New Internationalist; and Katherine Carlisle
Interior design: Verena Velten

CopperHouse is an imprint of Wood Lake Publishing, Inc. Wood Lake Publishing acknowledges the financial support of the Government of Canada, through the Canadian Book Fund program for its publishing activities. Wood Lake Publishing also acknowledges the financial support of the Province of British Columbia through the Book Publishing Tax Credit.

At Wood Lake Publishing, we practise what we publish, being guided by a concern for fairness, justice, and equal opportunity in all of our relationships with employees and customers. Wood Lake Publishing is committed to caring for the environment and all creation. Wood Lake Publishing recycles, reuses, and encourages readers to do the same. Resources are printed on 100% post-consumer recycled paper and more environmentally friendly groundwood papers (newsprint) whenever possible. A percentage of all profit is donated to charitable organizations.

How Maps Change Things was first published as an e-book by ODTmaps.com and New Internationalist.

LIBRARY AND ARCHIVES CANADA CATALOGUING IN PUBLICATION
Kaiser, Ward L., 1923-
    How maps change things : a conversation about the maps we
choose and the world we want / Ward L. Kaiser.
Issued also in electronic format.
ISBN 978-1-77064-566-0
    1. Cartography–History. 2. Cartography–Social aspects. I. Title.
GA203.K35 2012   912   C2012-907214-1

Library and Archives Canada Cataloguing in Publication
Kaiser, Ward L., 1923-
    How maps change things [electronic resource] : a conversation
about the maps we choose and the world we want / Ward L. Kaiser.

Electronic monograph in ebook format.
Issued also in print format.
ISBN 978-1-876998-08-0

    1. Cartography–History. 2. Cartography–Social aspects. I. Title.

GA203.K35 2012   912   C2012-907215-X

Print edition published by CopperHouse
An imprint of Wood Lake Publishing Inc.
9590 Jim Bailey Road, Kelowna, BC, Canada, V4V 1R2
www.woodlakebooks.com
250.766.2778

Printing 10 9 8 7 6 5 4 3 2 1
Printed in the United States of America by King Printing Co.

For

Gerhard Kremer (1512-1594)

and

Arno Peters (1916-2002)

Separated by centuries
but united in a common goal:

to map a better world

TABLE OF CONTENTS

Illustrations

A New Way of Seeing

"We are trying to create conditions like those in ancient Greece, which saw the flowering of ideas because people looked at the world with new eyes."

Neil Turok, Director of The Perimeter Centre, Waterloo, Ontario, welcoming Stephen Hawking

Young Neil Tyson kept talking even while he walked out of the Hayden Planetarium in Manhattan. He had just toured it for the first time. Part of the American Museum of Natural History, the Hayden regularly shows people things about space they could never even guess at. Small wonder this kid from the Bronx turned skeptical: based on what he had seen night after night with his own eyes, the sky they projected at the planetarium was just plain wrong.

"It's nothing but a hoax! The sky doesn't look like that!"

Did he let his doubting be the final word? No way! Neil got hold of a pair of binoculars so he could peer at the moon. When he was 14 he went to a youth camp in the Mojave Desert; there in the clear night air he saw stars he had never known from city streets. He took a course in astrophysics at the Bronx High School of Science. He pursued his education at Harvard, then at the University of Texas/Austin and at Columbia. Earning his Ph.D. didn't slow him down: he used telescopes over much of the world. He is now Astrophysicist of the American Museum and the Frederick P. Rose Director of the very planetarium that had so upset him as a youth. To many, his face

and name are familiar as the frequent and ever genial host of TV's NOVA or as writer of a monthly essay, "The Universe," in Natural History Magazine. To mark their esteem, the International Astrophysical Union even named an asteroid in his honor.

Shall we read this as a modern success story? *African-American youngster from poor neighborhood achieves world renown as space scientist.* If so, the account resembles a typical Horatio Alger narrative, mixing elements of high intelligence, strong determination, an independent spirit, community support, the ability to stay with a problem until it surrenders its secret. And maybe a generous helping of good luck, too.

At another level the story grows even more fascinating – though more straightforward. Everything hinges on Tyson's ability to look at things in a new way. To let go of the idea that what he had always "known" was absolutely right. To dream. To test ideas. To open himself to new possibilities. Without that openness of mind, Tyson's story would have dead-ended before it began.

That's what this book is about. Clearly, it's about maps – many maps – with special attention to the

Peters map of the world. At a deeper level it's about how we shape and use maps and how they in turn shape us. In short, it's about you and me and ways of seeing: how we see the world and therefore how we understand our place in it, how we connect to it and to all the people with whom we share this spaceship called Earth.

Such concerns have always been important; never have they been more central to the way we live on the planet. In the aftermath of September 11, 2001 and in the context of relentless technological shifts, runaway social change, and budget crises, talking about maps may seem impractical, even trivial. This book aims to persuade you otherwise. We contend, to be upfront about it, that in the world we are now entering maps belong front and center.

We offer no guarantee that this book will raise your IQ or your ability to steer an independent course or provide round-the-clock support for whatever goals you set. We don't promise you a Ph.D. or an important job. Nothing in these pages will by itself heal a wounded world. What we will be focusing on is the need and the possibility of gaining a new vantage point, a different perspective.

Welcome to a great and useful adventure – and even, possibly, to the satisfaction of making a difference!

CHAPTER 1

Framing Questions, Managing Answers

"The fascination of maps is found... in their inherent ambivalence and in our ability to tease out new meanings, hidden agendas, and contrasting world views from between the lines."

J. B. Harley

We all use maps as tools of exploration. But the most basic – and most surprising – discovery we can make with maps goes far beyond any factual information; it comes in that "Aha!" moment when we perceive that maps are loaded not just with data but with meaning. Maps speak to us at a deeper level and do more than we might guess.

To the untutored eye maps exist to provide specific information: where a city is located, how to get from place A to place B, what countries border a particular body of water. Straightforward, totally factual, utterly reliable.

If maps just did that, you wouldn't need this book. Neither would we need the dazzling variety of maps we now enjoy. What would there be to discuss? Wielding unquestioned power, the authority of the map would wipe out all differences of opinion.

In the real world, however, things are seldom that simple... or that bland. Maps stretch our minds, letting us see the world in ways we might never have imagined. Maps matter, not only because they pro-vide factual information but because they confront us with large questions and working answers, as we hope to show in the pages that follow. Maps also exist in a social context, which this chapter will illustrate.

Questions that Never Go Away

Among the most significant questions human beings ask - in addition to *Who am I?* – are *Where am I? How did I get here? Who or what is over there?* Through responding to just such questions we develop self-understanding and a sense of how we connect to the world and the people around us.

Politics, language, faith, philosophy, psychology, sociology, anthropology, economics, history, archae-ology, art, evolution, and astronomy all contribute to this understanding. So does the creation and study of maps – known as cartography (from the Greek *chartis,* map) - and its sibling, geography, meaning the study of the Earth (Greek *geo,* earth).

Just as maps are vastly more than tools for finding our way, actually tackling some of life's core questions, so this modest book boasts big ambitions. For many, it will provide a doorway into the per-haps arcane operation of how maps are made and the

Ceci n'est pas une pipe.

Was French artist René Magritte playing with our minds when he declared "This is not a pipe"? Come on, Mr. Magritte, what else would you call it?

Or was he saying that what he created is only a representation of a pipe and not the real thing at all? In what sense are maps interpretations to be read with understanding and care?

many ways we shape them and they shape us. It can be your key to understanding an important and especially controversial world map, the Peters. It may also, in ways difficult to measure but nonetheless real, contribute to your ever-changing perspective on who you are as a person and how you find your place and purpose in the world.

Let's turn to a specific example.

Assignment: Solve Our Energy Problem

Spring was bursting out in Washington, D.C.; day after day came alive with energizing beauty. Kwanzan and Yoshino cherry trees, given by the Japanese people, were exploding into bloom, harbingers of hope and global peace. Migratory birds were back; tourists strolled the boulevards between visits to the halls of Congress and museums. Workers from nearby offices soaked up the sun's rays as they took time for lunch on park benches.

Inside a federal building a group of powerful men gathered around a polished table. Under the guidance of Vice President Dick Cheney, they constituted the National Energy Policy Development Group, or "the Energy Task Force." With that title as a clue, we might assume the group would be concerned with the big picture: all aspects of energy supply and demand. That logic would gain support as we read a Presidential Memorandum of January 29, 2001, in which George W. Bush called on the group to develop "a national energy policy designed to help the private sector, and government at all levels, promote dependable, affordable, and environmentally sound production and distribution of energy for the future."

On the demand side, then, they might look at questions of

- conservation, including building insulation, low-energy appliances, and standards for vehicle efficiency.
- the power grid: how to meet demand peaks by shifting available supply.
- how adjustments to price might affect consumer demand.

- questions of safety and reliability in delivery systems.

Available evidence, however, suggests that this newly minted Task Force chose to focus solely on **supply**. Questions of demand didn't make it onto the agenda. (We say "available evidence" because, even though the Task Force met in the spring of 2001, its proceedings have never been made public.)

To focus on energy supply, then, one might assume they would survey a full range of options, including – for starters –

- nuclear
- wind power
- tidal
- hydroelectric
- geothermal
- solar
- coal
- oil
- natural gas
- agricultural sources such as corn, as used in ethanol
- extracting energy from waste
- plus any others their assembled brain-power might conceive.

For example, as a matter of speculation they might have sought expert advice on transforming vibrating air into electrical energy (precisely what we all do every day: the ear gathers sound waves that set up ripples in the fluid of the cochlea or "inner ear" that in turn pass over hair cells whose induced motion becomes electrical impulses; these in turn are carried to the brain and received as meaning and information). Who knows what creative possibilities they might come up with? In other words, what would it take to ramp up supply to meet growing demand?

Again based on the evidence, the team turned away from such a broad look; instead, they limited their focus to **oil**. Though they were called the Energy Task Force, they interpreted their task narrowly, and concerned themselves exclusively with *fossil-based fuels*.

With that decision made, the Task Force might logically turn its attention to America's major sources of oil, both existing and potential. These would include such places as

- Canada (which supplies more oil to the United States than any other country)
- Saudi Arabia
- Domestic sources, both offshore and land-based
- Nigeria
- Venezuela
- Iraq

That's where the third big surprise comes in: not only did the Task Force neglect demand to focus on supply, not only did they exclude all sources of energy other than crude oil pumped from under the ground, when they considered where oil was to be found they zeroed in on one country above all others: Iraq.

Maps Provide Perspective

Any task force looking at Iraq does well to use maps. Since they were meeting in Washington, D.C., this one had impressive resources at hand. The offices of the National Geographic Society, for example, stand only blocks away. The Federal Government certainly has a wealth of maps. A phone call or two would have brought a small mountain of cartographic resources to the table. Fig. 1-1 is an example.

Or the Task Force might have opted for some of the many maps available on the Internet. Fig. 1-2 provides an example.

Fig. 1-1 What information can you extract from this map: national borders, major cities and towns, distances, neighboring countries, position (coordinates), major rivers, lakes, access to the Persian Gulf...?

Source: United Nations. Map No.3835 Rev 5

Fig. 1-2 While this map agrees with Fig. 1-1 at many points, you will find important differences. See, for example, the border between Iraq and Saudi Arabia: This map adds a "de facto" line. Compare the selection of cities and towns. If your question has to do with commercial activity or terrain, which map might be more useful? How would a person get to Baghdad, the capital, from say Nukhayb (in the west) or Rayat (in the north)?

Source: sf.factmonster.com/atlas/country/iraq.html

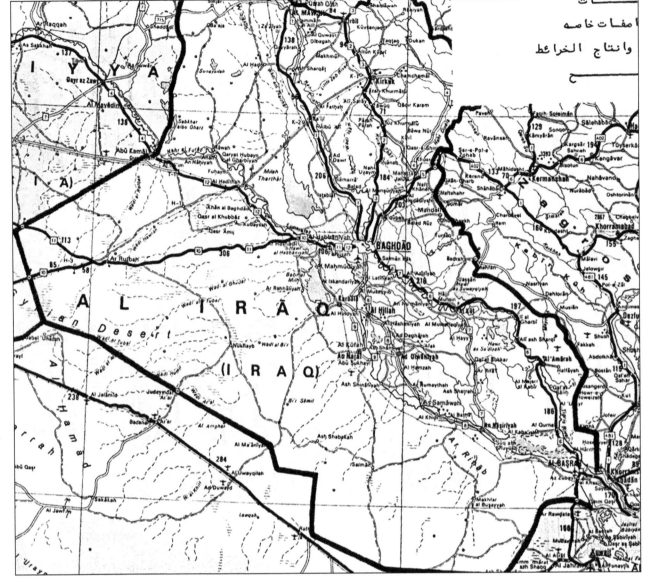

Fig. 1-3 This map, utilizing satellite data, adds information about highways (see the numbered boxes) and major and small airports (see symbols). Note also that it shows more lakes than Fig. 1-1 or 1-2. Can you get a clearer idea of the topography? Note the relative lack of towns on the western side of the country, the label Syrian Desert, plus the many "wadis" or seasonal river beds (usually dry, often steep-sided). Of these maps, which show regional capitals? The Tigris and Euphrates rivers?

Source: Atlas published in Kuwait.

If team members wanted a map from the area itself – often a useful idea – there were good ones to choose from. Fig. 1-3 shows a map of Iraq published and used in the Middle East.

In point of fact, the Task Force apparently rejected all existing resources in order to create a map of its own – see Fig. 1-4.

What would you say is the purpose of this map? General interest or specialized? Does it depict a nation-state along with its infrastructure, or is it a map of oil fields? Compare it to the other three maps of Iraq: how much does it tell us about topography, desert areas, where the cities, towns, roads, and airports are located? What information does it provide on the river systems that Iraq depends on? In bringing information about oil to the forefront while de-emphasizing population centers, does the map subtly suggest that Iraq functions primarily as a place to get oil? (Forget that it is home to 26 million people who live and love; forget its rich archaeological resources as the site of an ancient civilization.) One may criticize the creators of this map... or one can accept that they

Iraqi Oilfields and Exploration Blocks

were simply doing what all mapmakers do: selecting what suited their purpose.

These four maps well illustrate a central, essential fact about all maps: each is selective. Maps select and frame a particular piece of geography – that much is obvious. More important, and more surprising, is this: every map frames a question, which then becomes the question. Every map then responds to its own question by selecting those data deemed relevant. No map shows everything; what is selected reveals the mapmaker's mindset or purpose. To apply that general principle to the example before us, if the Task Force on Energy had been assigned the task of securing Iraq's oil for America, Fig. 1-4 would be a very good map. Rejecting all clutter, it single-mindedly draws attention to what supports that purpose.

> **"I am saddened that it is politically inconvenient to acknowledge what everyone knows: the Iraq war is largely about oil".**
>
> Alan Greenspan, former chairman, U.S. Federal Reserve, in September 2007.
>
> Greenspan and others took several years to reach their conclusion. What if they - or all of us - had analyzed the Task Force map (as you have just done) before the invasion? What would have happened if the Task Force map had been widely circulated? Might the course of history have shifted in another direction?

But maps do more than reflect intention. They also *create* a mindset. They nudge us toward a particular view of reality. What maps say to us, or show us - the "answer" they set forth to the question they frame -

exerts a powerful influence on our perception, especially since maps still carry an aura of being reliable and bias-free.

What we're saying is this: *maps are verbs*. They may seem to be tactile objects, documents we can handle or fold – *nouns* - but don't be fooled. In persuasively framing questions and selectively supplying answers they act; they initiate; they function as agents. This is "the power of maps."[1] Just as a map of the United States that highlights the network of Interstate highways (while suppressing, say, historic sites or economic realities or elevation or voting patterns) sends one message, so each of these maps depicting the same chunk of real estate carries its own message. A map of Iraq that says, in effect, "Here is a rich and tappable source of oil," will influence those who see it.[2]

So I assert that buying a map is not an innocent choice. Why? Because maps have agendas. By the

Everybody has an opinion on the Iraq War – soldiers and bloggers and Jane Q. Citizen. How many know how a particular map changed things, building up a mindset that changed the future both of Iraq and every nation that joined the coalition for regime change in Iraq?

way they frame and answer questions they may quietly shift the way we perceive reality. It's no exaggeration to say that those who control our maps help shape the outcome of our lives.

To summarize: maps are always more than they seem. They may look like squiggles – however accurate and useful – on a page, but they actually *mirror* a purpose and *shape* a perspective. Both of these actions occur simultaneously. The next chapter will explore this more fully.

A Scenario

If I were writing a play or a movie script to show how maps reframe our questions, then customize answers, I probably couldn't do better than to give center stage to the Energy Task Force. I'd show them as intelligent, top-of-their-form, Type-A personalities. They know how to run a business; they have an impressive history of getting results.

They're human as well. Very much aware of waistlines (expanding) and time horizons (shrinking). Guys who know what tension is, and ruptured relationships. At least as likeable – and perhaps as flawed – as the rest of us.

What they don't have is map skills. Not one is a cartographer (cartographers never get appointed to task forces like this). So members of the team commend the advance work done for their meeting. Especially the map of Iraq. "It's clear, it's factual, it's

a solid basis for developing our recommendations!" one of them explains, enthusiasm in his voice. One person with a bit of exposure to map construction might even point out that its angles, sizes, shapes, and distances speak to the professional expertise behind it.

Just off stage, semi-seen by the audience, stands a stash of maps offering other perspectives on Iraq: topographic, commercial, water supply, you name it. But the team is content with what they have been given. One member, channeling Ira Gershwin, sings, "Who could ask for anything more?"

So they are seduced by the appearance of objectivity, unaware that the map they have is not a map of Iraq at all but a map of oil in Iraq, unaware that the focus of their inquiry has shifted. The map before them is changing things. And they don't even know it.

CHAPTER 2

Maps Send Messages

"...there can be no such thing as a truly objective map. All maps are to a greater or lesser extent, and always have been, propaganda." [1]

Simon Winchester, author of *The Map that Changed the World*

Every map packages an idea. When there's a match between what a map sets before us and our expectations, we may hardly notice that the map is, after all, not the real world but only one way of looking at it. When a map diverges from what we expect, showing us another way of viewing reality, we may be stimulated, even gaining new insight, or puzzled, disturbed, angry, or confused. The map may make us so uncomfortable that we simply reject it.

What's a Map For?

Ask the right questions. If you're not sure what the right questions are, ask anyway. Keep an open mind. These are key to gaining benefit from any map. Our worldview may be stretched and enriched with new ideas. Here's a case in point: near the end of a course on mapping, I may invite students to comment generally about what maps are and do. Here are a few responses, gleaned from a college course, a continuing education program for high school teachers, and a church mission study on maps. You may find the range of responses revealing.

What's a map? Obviously, it's a document; in the past typically a sheet of paper, increasingly today an image on a screen showing the world or some part of it. What does it do? It provides useful information such as names of places, distances, types of terrain. In many cases it shows us how to get where we want to go.

If such a response seems obvious, no need to consider it any further, please read on. You may be in for a surprise.

Building on their recent exposure to a variety of map images, participants responded with analytical skill and their personal feelings:

- a neat way to get certain kinds of information
- highly political statements
- enrichment for following the news or the Olympics
- a prelude to prayer
- propaganda in disguise
- a way of connecting to people I've never met
- it's humbling - it puts us in our place.
- helps students "really connect" with their studies in literature, language, history, math and science
- a helpful corrective to some of our mistaken ideas
- a way to separate "us" from "them"
- a device for telling me things I could never guess at
- a way to get answers to questions like How far? How big? Where?
- a way to demonstrate "We're all in this together"
- some maps make me almost cry because they bring home what's going on.
- a reminder of the richness and variety of the world
- every time I see a map of our country it's like falling in love.
- what a wonderful world!

Those diverse responses functioned as windows into students' thought processes. The statements were never intended to be definitive, just markers along a journey of discovery. At one time these students, like many of us, may have seen maps simply as factual documents – a map is a map is a map – but through their exposure they began to read them at a deeper level. They understood the "messages" that lay behind the map and that were there all the time.

Now let's turn to a particular case.

Deciphering the Mercator Message

A prime example of maps sending messages is the familiar Mercator. We'll have more to say about it later, but right now let's simply state that it's a perfect map in some ways but perfectly awful in others.

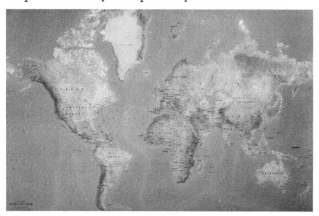

The Mercator projection

Consider the positive aspect first. When used for its original purpose – navigating – the Mercator asserts, as clearly as any projection[2] system can, that the world is predictable, measurable, reliable. Even manageable. Angles on the map accurately represent angles on the surface of the Earth. It makes no difference whether you measure those angles along the equator or in high latitudes, at the edges of the map or near its visual center; if you're a sailor you can trust your ship and your life to the answer. Moreover, what was true in 1569, when the map was devised, remains

equally true today; technology has not outdated it. The world, like the map, is – in this respect at least – user-friendly. As surely as daylight first appears on the eastern horizon, this map can be trusted.

On the other hand, when a world map on the Mercator projection is used – as happens far too often – as a wall decoration, a symbol standing for the whole world, to promote a product, for classroom teaching, or as context for negotiating how the people of the world will live together, it sends the wrong message.

Why? Because it grossly enlarges some parts of the world and seriously diminishes the size of others. Given that India is about three times the size of Scandinavia, a map that accurately shows that ratio might be said to be fair, free of bias, value-neutral. But the Mercator makes the larger area look smaller. Similarly, Russia shows up larger than all of Africa:

> **Map Fact**
>
> Conformal: the name given to a projection that maintains correct angles and shapes at any given point. The Mercator is the example most commonly cited.

that is, the reality on paper does violence to the reality on the ground since Africa is close to twice the size of Russia. The essential point is: Where size is concerned, better not count on a world map on the Mercator projection.

There's more: the way the Mercator is typically presented drops the equator low on the page, giving about two-thirds of the map's space to the northern hemisphere and placing, say, Europe near the visual center of the map. What message does that send? How about these:

North rules!

Europeans are the center of the world.

Some of us are big and powerful.

"South" is the graphic equivalent of the segregated back of the bus in the days before Rosa Parks initiated the Montgomery bus boycott and so created a critical moment in the U.S. civil rights movement.

Some people count for more than others.

"It's all about us!"

In graphic terms it sets before the viewer a contrast: This is *US* – This is you.

Still, let's be fair to Gerhard Kremer, the map's creator. In situating Germany, the country to which he had fled as a refugee, near the visual center of his map, he was following age-old precedent. As humans we have a pervasive – universal is not too strong a word – tendency to place ourselves at the center of things. (How could it be otherwise? Stand somewhere... gaze around while you turn full circle, then in your imagination project your view outward, farther and farther until it takes in the whole world. Don't you get a virtual map centered on yourself?) So here's an approach to reading map messages: on any world map – often on other maps as well – pay attention to the visual center. What you find there offers one clue to "important" and to "home." Specifically, to the home base of the mapmaker and what places really count. Often, as with the Mercator, the two merge into one.

Fig. 2-1 Greenland vs. Africa on this use of the Mercator Projection.
Source: www.ODTmaps.com

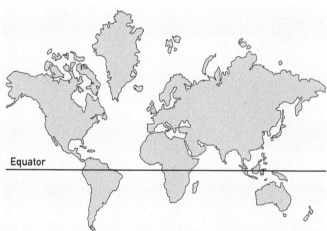

Fig. 2-2 Notice the low placement of the equator on this map in the Mercator Projection.
Source: www.ODTmaps.com

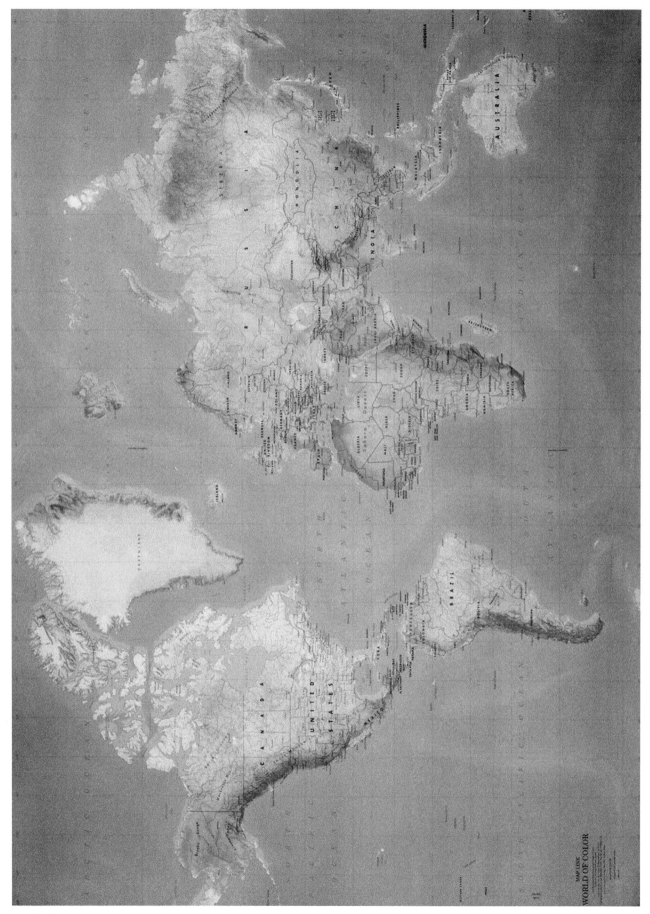

Fig. 2-3 World map using the Mercator Projection

Source: www.ODTmaps.com

Little Town, Big Message

Maps can do many things. For one, they can legitimately depict a world centered on the viewer, as in the process just described. Within that category are maps known as equidistant. As a projection system this has the playful capacity to be centered anywhere you choose. Thus, if you believe "all roads lead to Rome," you can set the center there: you look out on the world as if that is where you are standing. If you hold, as the Human Genome project declares, that humans first appeared in Africa, you might center it there and plot the probable migrations outward. Fig. 2-4 shows the whole world as if the tiny town of Beamsville, Ontario, Canada (population 10,000) stands at the center. Suddenly, it's as if the world revolves around this tiny community!

Does the map send a message? Very clearly! It says that Beamsville, too small to appear on many maps, sometimes subsumed under its town name of Lincoln, counts. It has as much right as any other place to consider itself the center of our planet! Reinforcing that impression are the concentric rings, designating distance to Beamsville in kilometers. Is it a good map? One might find fault with it: it distorts shape (note Australia especially). As for distance, only distance to/from Beamsville is correct; all others will be wrong. Yet the map is perfect – for its purpose.

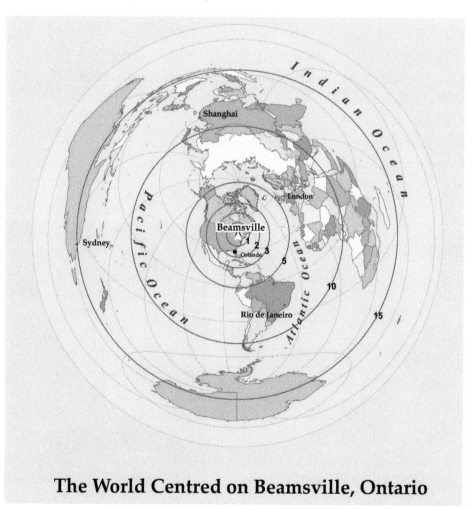

The World Centred on Beamsville, Ontario

> **Map Fact**
>
> Equidistant: Any map that accurately shows straightline distance from a central point. Fig. 2-4 is an equidistant map. The Mercator, on the other hand, sacrifices accurate distance over the entire surface (except along the equator) in the interest of correct angles. The Peters and the Hobo-Dyer (which we will introduce in later chapters) sacrifice measurable distance for the sake of getting size right.

Fig. 2-4 Shows the whole world as if the tiny town of Beamsville, Ontario (population 10,000) stands at the center of everything. Suddenly, the world revolves around this tiny community!

See full size map on page 160

Source: www.ODTmaps.com © Len Guelke, from the author's collection.

Fuller Makes His Message Clear

When futurist Buckminster Fuller launched his map, which he labeled the Dymaxion World Map or Dymaxion Airocean World, he articulated its messages. Since his flat map (Fig. 2-5) may be assembled into a multisided "globe" in several different ways, he pointed to the map's several meanings:

> ...one of these pictures is... the One Ocean World, fringed by the shoreline fragments... It discloses the relative vastness of the Pacific and emphasizes that ocean's longest axis, from Cape Horn to the Aleutians. Oriented about the Antarctic, the waters of the Indian and Atlantic Oceans open out directly from the Pacific as lesser gulfs of the one ocean... compare the impressions derived from looking first at the one-continent arrangement and then at the one-ocean assembly. Turning away and reporting his [sic] impressions... [the reader] would be inclined to testify that these maps were composed of different components; that the one-continent map was composed of seventy-five per cent dry land area, that the one-ocean map was comprised of ninety per cent water area.[3]

Either way the message is striking: the world is a vast ocean (punctuated by chunks of land, outcroppings of the world's one continent) or one large land mass (flooded in places by the surrounding sea).

Fuller also saw his invention as a long-needed corrective to an outdated perspective on the world. He spoke of

> ...our prevailing public ignorance of dynamic Air-ocean geography, we being as yet blinded by... Mercator projection vendors, and by the historical east-west orientation inertia. It is hoped that a new north-south dynamic world orientation will be aided by the Dymaxion Airocean World.[4]

Fuller was also known for his now-famous expression, "We're all astronauts aboard a little spaceship, called Earth." His map, foldable into a 20-sided globe (below, right), reinforced that idea.

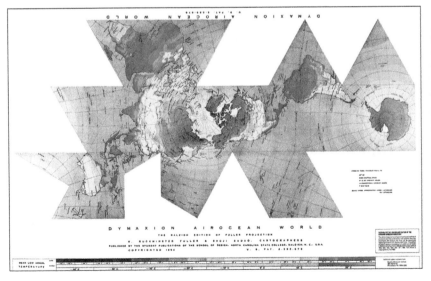

Fig. 2-5 Dymaxion World Map. See full size map on page 157

Source: BFI

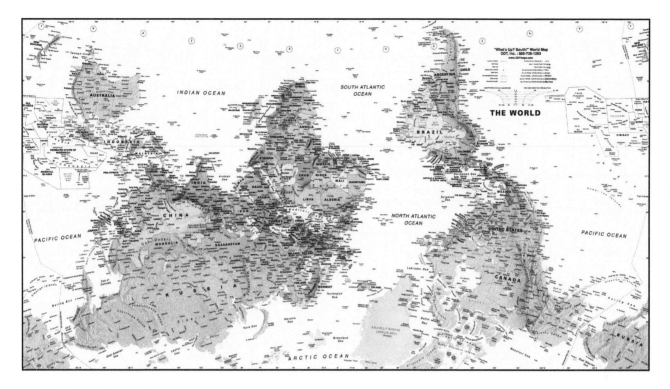

Fig. 2-6 What's Up? South! Map. See full size map on page 153

Source: www.ODTmaps.com

Message: What's Up?

With the stunning clarity of an alarm clock comes the message of the What's Up? South! Map (Fig. 2-6). The interpretation printed beneath the map opens with the provocative question, "Who ever said North must be on top?"

First, let's be clear about one thing: it's convenient to keep north consistently at the top of a map: you don't have to check the compass symbol printed on the map every time. But north doesn't have to be given pride of place. In some earlier centuries, east got top billing, and our continuing use of terms like "let's get oriented" reflects that practice. ("Orient" refers to east; get east right and everything else falls into place.) One of Mercator's own maps – of the British Isles – has west uppermost. South has often been set at the top; that practice is sometimes still followed by organizations that cater to motorists, including the American and Canadian Automobile Associations, books like *Along Interstate 75* and some British road maps. So even for drivers heading to Florida, south may be "on top." Some software allows you to print strip maps either "up" or "down" – whatever "feels" right, based on the direction you are going. All of this points to a fundamental fact: what goes "up" owes more to convention and personal convenience than to any requirements of cartographic science.

That's why this next map (Fig. 2-7A), known as the Hobo-Dyer, is both startling and refreshing: it shows us another way to see the world. Actually two different ways.

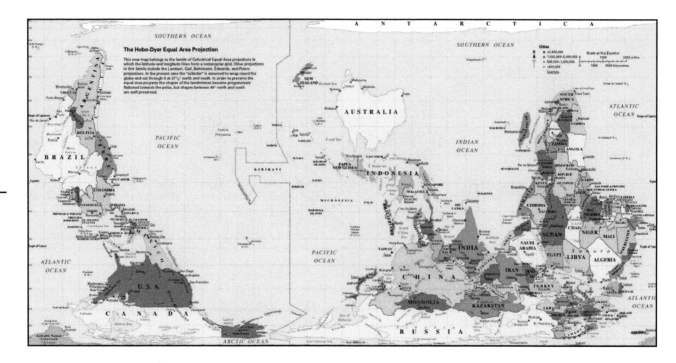

Fig. 2-7A Hobo-Dyer south up, Australia-centered Map. See full size map on page 155
Source: www.ODTmaps.com

The first is, of course, the south-up perspective. Maps with this orientation are particularly well known in Australia and New Zealand. The text printed on Stuart McArthur's map (Fig. 2-8B), another south on top map, is instructive:

> ...the first step in the long-overdue crusade to elevate our glorious but neglected nation from the gloomy depths of anonymity in the world power struggle to its rightful position – towering over its Northern neighbors, reigning splendidly at the helm of the universe. Never again to suffer the perpetual onslaught of Down Under jokes – implications from Northern nations that the height of a country's prestige is determined by its equivalent spatial location on a conventional map of the world... Finally, South emerges on top.[5]

McArthur's map soon resonated with Australians and has been used for educational and humorous purposes in the Northern Hemisphere as well.[6]

For a second startling insight to come from the Hobo-Dyer image, compare Figures 2-7A and 2-7B. Though the images seem sharply different, they have only two differences, what goes at the top and what goes at the center. The traditional presentation (Fig. 2-7B) places the Prime or zero meridian (about which we'll have more to say later), running through Greenwich, England, at the center. Africa and Europe claim a certain priority of attention. An alternate way to slice the map centers it on the Pacific. Many viewers immediately get the point: Fig 2-7A not only promotes Australia's visibility but dramatically emphasizes how much of the world is water. Such a message impacts us all, given that about 70 percent of our planet's surface is water (though only three percent is fit to drink).

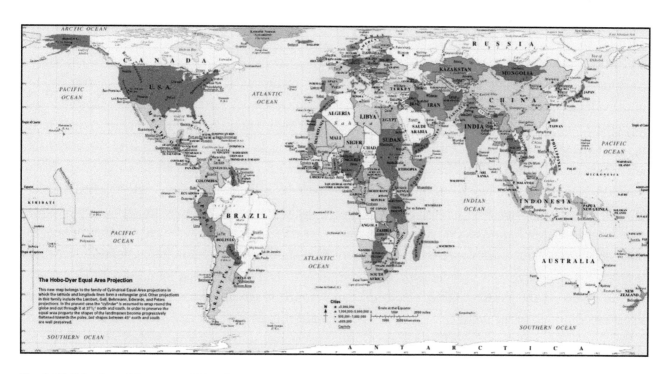

Fig. 2-7B Hobo-Dyer Africa-centered Map. See full size map on page 156

Source: www.ODTmaps.com

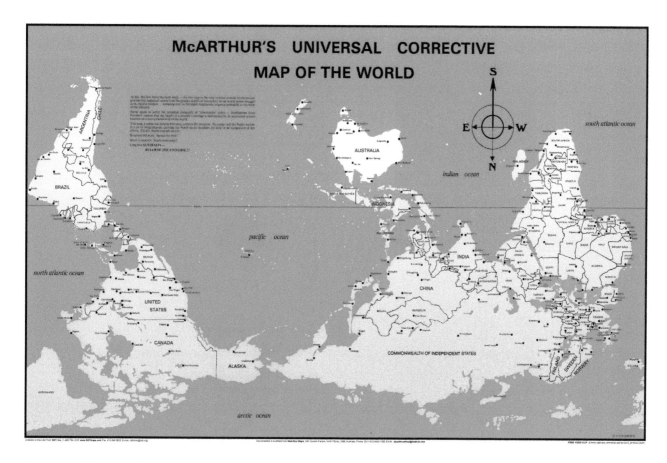

Fig. 2-8 In 1979, Stuart McArthur´s tongue-in-cheek humor ushered us into the world of modern south-on-top maps, which had lapsed since the late 1500s. After the invention of the magnetic compass, North-on-top soon became the accepted convention, and eventually - the rule. See full size map on page 159

Source: www.ODTmaps.com

Where a "Simple" Decision Alters Political Perceptions

Imagine a light source inside an earth globe. Then imagine wrapping a single sheet of paper around the globe, touching it, say, at the equator. Turn on the light; the land masses and lines visible on the globe are now projected onto the flat sheet of paper, hence the term "projection" (see Fig. 2 - 9).

Since the wrap-around sheet forms a cylinder, this type of projection is known as cylindrical. In actual practice, it is not created that way, but the illustration is an excellent metaphor for the mathematical mapping of points on the globe to plotted points on the flat sheet. But back to the cylinder...

One more task remains: the cylinder must be cut to show it as a flat sheet. Where is not dictated by the globe, which has no slice markings; it is the mapmaker's free choice. And that choice influences our perceptions, as we've just seen in the case of the Hobo-Dyer.

In reality, what seems like an innocent and purely technical choice by a disinterested mapmaker may have implications for international relations. For example, to locate North America at the visual center (which might be desired in some cases), Russia must be split. Some mapmakers rejected that approach, preferring to show the vast sweep of Russia – or the Soviet Union in its day – uninterrupted. But that choice may translate into a political statement, says cartographic analyst John Snyder. "Contrary to wanting to de-emphasize Russia's importance, there were rumors that its large red size on some U.S.-made maps in Mercator projection – with Russia not split – was to help justify the military desire for U.S. military build-up."[7]

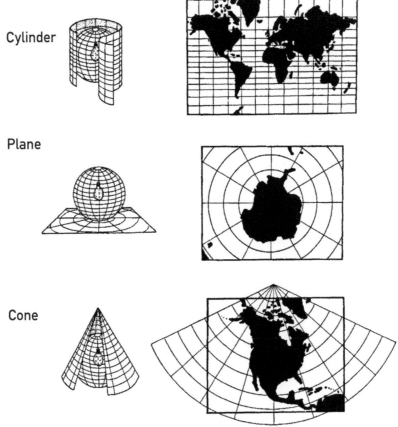

Cylinder

Plane

Cone

Fig. 2-9 You can project the globe onto anything that will roll out flat when you cut it. Planes, cones and cylinders are most common. Note the effect each has on the pattern made by the lines of latitude and longitude.

A Subtle Message from a Border

A widespread impression holds that Canada and the United States are so interconnected they are destined to draw closer and closer together. Perhaps even, one day, join. In spite of the current emphasis on "walling off" its borders to keep suspicious characters out, the United States historically shares with Canada "the world's longest undefended border." The two countries have long been each other's best customer. They speak the same dominant language, watch the same movies, laugh at the same jokes. Thousands of people call one of the countries home while working in the other. Uncounted numbers have family and friends in the other nation; according to State Department estimates, fully 300,000 people cross the border on any average day.

From this perspective, the boundary is a costly nuisance. In an ideal world it would be downgraded. The two countries would be integrated; common defense, free trade and the unimpeded flow of such resources as water, oil, lumber, and labor would wipe out its drawbacks.

The map, however, speaks of a realm of meaning just beneath the surface. It points to two different ways of living. The simple fact that the line exists at all – in spite of all the familial, cultural, economic, and political pressures to send it to the trash bin – is evidence of another reality. A group of visionaries in the 1860s chose, as *The Times* of London saw it, "to raise a barrier of law and moral force extending near three thousand miles between [Canada] and the most powerful and aggressive state in the New World." Today James Laxer looks at a map and says, "The Canada-U.S. border draws a line between two societies with strikingly different views on key contemporary societal questions. On guns and capital punishment, the environment and health care, war and peace, Canada and the U.S. march to different drummers."[8]

On Canada Day 2009 the *New York Times* invited some Canadians living south of the border to speak about how they perceived the meaning of the border. Some of their responses highlight the differences Laxer mentions, and which the map highlights. Whimsically yet with bite, Tim Long writes from sunny California about snow. "...to my Canadian eye, American snow is like American health care: sporadic, unreliable and distributed unevenly among the population."

Maps – like the lines across them marking a seemingly simple boundary – open up meanings beyond anything a casual tourist discerns.

For Whom the Bill Tolls

If boundary lines speak of diverging attitudes, so their absence can make a big difference. For this we turn to Florida, and to highway planning.

The goal seemed simple enough: build an interchange from Interstate 4 to Walt Disney World. Three principal entities developed a plan: Disney, and Orange and Osceola counties. A map of the area was provided: (Fig. 2-10).

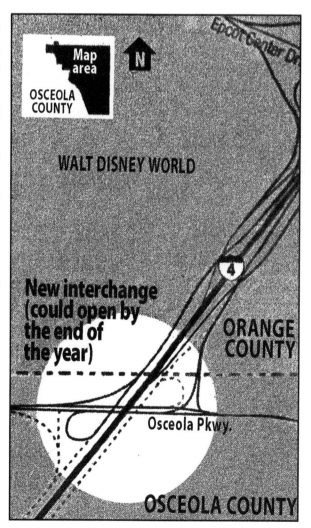

Fig. 2-10 Interchange in Orlando area
Source: Courtesy of the Orlando Sentinel

So far, so good. On and off ramps were designed, a schedule was adopted, a budget set up. Agreements were signed. Then came the bombshell: Orange County taxpayers were handed a bill of $53 million ... for an interchange in another county. Osceola County officials were later quoted as saying, "Cheapest interchange we ever got." In effect they picked up a free benefit: the interchange brought traffic, and therefore tolls and business, their way. In the controversy that ensued the facts came out: the engineers' map, *which did not show county lines*, made it seem that Osceola was just a minor player.

Was the misinformation intended? You don't have to believe that to get the point: the map had budget implications that taxpayers in Orange County will continue to pay for, for years.

From Fuzzy Maps to Persistent Problems

Like sheet glass hit by a heavy rock, peace suddenly shattered in Caledonia, a Southern Ontario town. It was early in 2006. On one side were First Nations people; on the other, commercial developers and their supporters. From February's freeze to summer's heat the issue was who did 40 hectares (about 99 acres) of land belong to: the native peoples or the newcomers? A housing development was in its early stages; builders and potential buyers believed they had rights; First Nations representatives challenged that. Over time the dispute heated up, with blockaded roads, insults being hurled, non-stop police presence, occasional acts of violence, people's lives being disrupted. Even the U.S. Border Patrol decided things were serious enough to take the unusual step of dispatching agents into another country's sovereign territory.

Outside observers might wonder what the fuss was all about. Just look at the deeds, they might say. Check official records of land transfers, get the legal facts and let that settle things.

"Things" didn't yield to simple solutions, however. History and maps played a role fully as important as legal deeds.

The story starts as the American colonies split from Britain. Thousands of people – known in the United States as Tories or those who had chosen the wrong side in the war, but who were honored in Canada as United Empire Loyalists – migrated north in the years 1775-1783. Upper Canada's Governor wanted land for them to settle on, as well as to provide for indigenous Six Nations people who had rallied to the British side in the American Revolution. Through Colonel John Butler, he procured land from the resident Mississauga Tribe.

But neither the Governor nor Butler knew exactly how much land they had acquired. Neither did the tribal chiefs who granted it. Why? Because the land had never been accurately mapped and measured. Efforts to define it soon ran into trouble as landmarks and boundaries, stated in words, left surveyors scratching their heads.

The story continues on October 25, 1784, when the Governor, Sir Frederick Haldimand, turned over to the Mohawks what may variously be called the Grand River Tract, the Haldimand Tract, Mohawk Territory, or Six Nations Land. This first grant of Crown lands in the history of Upper Canada (now Ontario), it encompassed an area extending

Source: Courtesy of **Expositor**, Brantford, Ontario

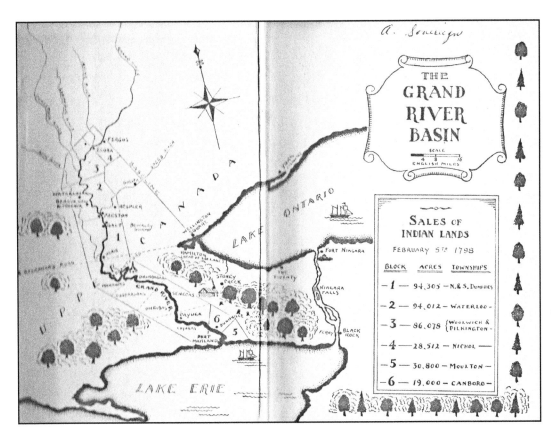

THE
GRAND
RIVER
BASIN
SCALE
4   8   76
ENGLISH MILES

SALES OF
INDIAN LANDS
FEBRUARY 5TH 1798

BLOCK   ACRES   TOWNSHIPS

—1— 94,305 — N.&S. DUMFRIES

—2— 94,012 — WATERLOO —

—3— 86,078 {WOOLWICH & PILKINGTON —

—4— 28,512 — NICHOL —

—5— 30,800 — MOULTON —

—6— 19,000 — CANBORO —

Fig. 2-11 Hand-drawn, thought to date from the 1930s but based on earlier sources.
Source: Collection of the author.

"Six Miles deep from each Side of the River beginning at Lake Erie, & extending in that Proportion to [its]… Head."[9]

The devil, as often, was in the details. Where did the Grand River begin? No one knew. (Actually, its source is in an extensive swamp, so where in that morass do you start measuring your six miles from?) Further, do you measure from the exact center of the river or from each bank? *When* do you measure: in the spring, when the river is in flood or when it dries to a summer trickle? What about tributaries: were they to be considered part of the river, or not? Say that the general direction of the river is south, do you measure six miles consistently east-west or, wherever the river bends, apply the rule perpendicular to the riverbed? Map-making efforts in this case, no matter how well intentioned and careful, left a lot to be desired: they were imprecise and subject to various interpretations.

By 1791 a new governor, John Graves Simcoe, ordered a survey of the entire province. In the process came a startling announcement: the land originally purchased from the Mississaugas did not extend as far as the settlers supposed. Early surveyors thought they had reached the Thames River when they got to what is today the Conestogo River, so the early calculations were confusing. The result was a serious question: did the First Nations hold technical and legal title to thousands of acres more than anyone had assumed?

Creating further uncertainty, the Simcoe survey map – for reasons unexplained – stopped short of the source of the Grand River. On the one hand Chief Thayendanagea of the Mohawks, also known as Joseph Brant, clearly believed that the lands along the Grand were "absolutely our own;"[10] on the other, if that opinion had been widely shared history would be very different.

While our story here is unique, it needs to be seen as symptomatic of a larger issue stretching across North America. Much of British Columbia's vast territory – some say 100 percent![11] – is subject to continuing land claims. Tanya Talaga, a reporter specializing in government affairs, summarized her findings July 17, 2012: "At any given moment, there are nearly 500 government lawyers challenging aboriginal rights in Canada." In 1946, the U.S. set up the Indian Land Claims Commission, but after 60+ years it still has work to do.

Agreements on fishing and hunting rights, governance and taxation as well as land transfers were often based on treaties and maps that – certainly by today's standards – were less than reliable. By and large, negotiations over treaty rights have been fairly amicable though painfully, frustratingly slow. Yet enough claims like the one in Caledonia are still unsettled to give cartographers, lawyers, politicians, justice activists, and negotiators steady work for years to come. No one intentionally falsified their maps but, clearly, both maps and the absence of maps are at the heart of the continuing controversy.

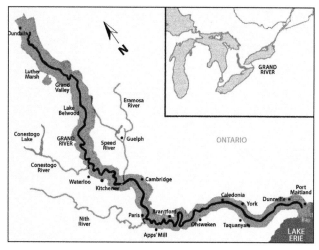

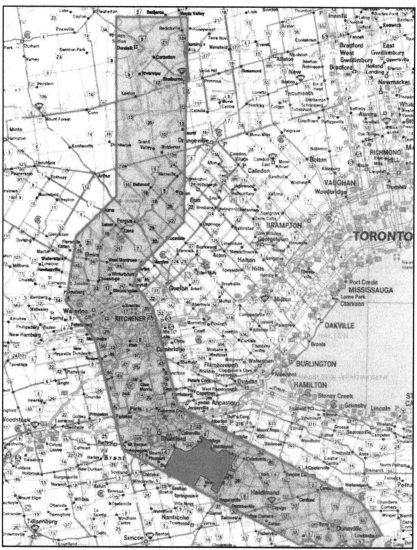

Fig. 2-12 Shows the benefits of more advanced mapping techniques: it is more accurate in scale and direction, for example. The shaded area represents one interpretation of the six-miles-from-the-river grant of land to the Six Nations.
Source: Collection of the author.

Fig. 2-13 Part of the Six Nations' effort to assert their claim. The shaded portion shows land thought to have been granted by Haldimand October 25, 1784; the block shaded in red/pink (center-bottom) shows current Six Nations land. This colored area measures about 46,500 acres, or 4.9 percent of the original grant.
Source: Reclamationinfo.com

Another Way of Locating "Up"

To raise the question of what goes "up" on a map is to invite a serious look at how North American indigenous people did mapping. Their experience has something important to add to the conversation. But first it is important to gain some background, to develop at least a basic rapport with their context.

Few maps drawn centuries ago by the first North Americans have come down to us.[12] As a result, some tentativeness, even considerable humility, is in order whenever we as outsiders deal with related subjects.

Still, some assertions can be attempted. Indeed, *should* be attempted, for deliberately to ignore indigenous people's spatial communications would be to demean their culture and impoverish our understanding. Let's start by asserting that North American natives created no maps of foreign lands. That sharply contrasts with European practice: explorers and conquistadors were constantly documenting their territorial ambitions with maps.

Yet European and American explorers often relied on mapping information supplied by indigenous people. Samuel de Champlain in the New France of the early 17th century; David Thompson, whom some consider the greatest geographer of all time and who almost single-handedly mapped the vast interior of British North America, including Montana and the Columbia River region as well as today's Canadian West; and Lewis and Clark, with their 1810 history-making map of the American West – these among others acknowledge their debt.

Further, the original residents of the Americas had no need of what we call cadastral maps – maps with boundary lines to support ownership or taxation status. With a different understanding of "this land is *my* land" than Europeans brought, they neither set up fences nor laid down lines on maps separating "ours" from "yours" or "theirs." We can also state, in broad terms, that the "maps" they did create were highly functional, showing the way to good hunting or fishing or where a cave might provide shelter in a time of storm. Toward this goal they relied on song and story as well as visual information. (Imagine, if you will, "speaking" a map or "singing" a map rather than "looking at" a map.)[13] Where modern maps count on measured, verifiable data, early indigenous efforts sometimes referenced the spirit world as part of the human experience. To our Western mind with its linear thinking and dependence on documents, the oral tradition may seem strange, foreign, unreliable... yet its value was strikingly affirmed by the Supreme Court of Canada in a landmark 1997 ruling on First Nations land claims.[14] Other times, as among the Inuit, stone markers provided information that in a literate culture might be conveyed through maps and written stories. Because they function as geographic markers *and also say*, in effect, to travelers in a harsh and lonely environment, "You are not alone," they blur the line between "geography" and "relationships," between demonstrable fact and perception, and finally between "maps" and "meaning."[15] It becomes crucial to look beyond the physical object to ask about significance and even, sometimes, about the mysterious. And that is true not only of Inuit markers but of any map.

Stone markers like this, known as Inuksuk (s.) or Inuksuit (pl.), served to provide information or guidance among the Inuit in Canada's far north and in Alaska. They also sent a message, "You are not alone. Others have come this way ." The 2010 Winter Olympics in Vancouver popularized the symbol. One Inuksuk holds an honored place in Canada's Parliament in Ottawa.
Source: www.seethewestend.com

Now to the question of how they oriented their maps. Did North get to the top? Interestingly, pictographs created before the European conquest suggest a whole other way of determining what goes "up" on the map.

In contrast to then-developing European practice, the top of the drawing could never be assumed to represent North. Nor South nor East or West. Instead of a compass direction, these mapmakers may be said to have thought outside the compass box, setting "upstream" – that is, higher ground – at the top of the map.

Does this strike you as surprising or strange? If so, it demonstrates how fully most of us have internalized the dominant Western system. It may be worth asking why we are prone to look down on the practice of indigenous peoples – in the Americas, the South Seas, Australia, wherever. Is it inferior, or just different? From another perspective, defining map orientation by the direction streams flow may have more in common with current modes of thought than we might suppose. A French friend once sketched for me a map of the area drained by the Seine River, from Paris to its mouth at Le Havre. He put Paris at the top of the page and finished with the English Channel at the bottom. In choosing to show the river flowing "down" on the page, he reversed the orientation maps typically follow. Interestingly, his choice conforms to the traditional naming of France's administrative regions or *départements* – the area around the river mouth is known as the Lower Seine (*Seine Inférieure*) – no matter that it is farther north. And sailors on the St. Lawrence-Great Lakes system speak of "going up" even when they are heading west or south, so long as they are moving against the current. In a similar way, bioregional mapping may emphasize a "watershed" approach, making clear that what is dumped upstream flows downstream.

Maps that look "primitive" or "unsophisticated" – lacking many features we associate with contemporary, color-coded, computer-based cartography – can yet be fully functional. That is, such proto-maps were doing exactly what they were supposed to do. And they were sending their own important message: there are more ways to create a map than most of us have ever imagined.

Fig. 2-14 North American Indian Pictograph

This pictograph illustrates how indigenous tribes in North America used highly functional communication systems. Map users today expect to be told where North is; those for whom pictographs/maps like this were drawn found upstream/downstream and locating good hunting areas more to the point.

Source: Public Domain

Mapping for Peace – Without a Map

Now let's time-travel to Camp David near Washington, D.C. The year is 2000. We look in on negotiations that we hope, as the whole world fervently hopes, will bring peace to the Middle East. Or at least something significantly closer to peace between Israelis and Palestinians. One common perception of the outcome is that U.S. President Bill Clinton and Israeli Prime Minister Ehud Barak have come up with a generous offer, which the Palestinians perversely reject (Fig. 15B).

Noam Chomsky begs to differ: "There is a simple way to evaluate these claims: present a map of the territorial settlement proposed. No map has been found in US media or journals, apart from scholarly sources and the dissident literature." Evidently with good reason, certainly with major result: "A look at the maps reveals that the Clinton-Barak offer virtually divided the West Bank into three cantons, effectively separated from one another by two salients consisting of expansive Jewish settlement and infrastructure developments."

Dennis Ross, a Fellow at the Washington Institute for Near East Policy and chief negotiator for the U.S. in the Camp David process, agrees with Chomsky to this extent: "we actually never put a map on the table... We were leaving it to the parties to... develop the maps."[16]

Develop their own maps – that is precisely what they did. In effect, the map shown as Fig. 2-15A makes clear what mere words in a news release could hardly show: that the proposed solution would, according to one understanding, subject Palestinians to a permanent state of isolation and dependency. In fact, "In Israel, maps did appear in the mainstream press, and the proposals are commonly described as modeled on South Africa's Bantustans of forty years ago"[17] (with the same oppressive intention and result).

Would the right maps, widely circulated, actually change what people thought – and still think – of Camp David 2000? Chomsky, a leading analyst of current affairs, clearly says Yes![18]

We started this chapter asserting that maps come with purposes, perspectives, and messages. Sometimes these messages are blatant and at other times hidden. We looked at the Mercator, locale-centered maps, as well as Buckminster Fuller's Dymaxion, and maps with different centering and different compass orientations (say, south on top). Borders shape and reflect political perspectives. And mapping from indigenous cultures represents a challenge to Western-based modes of map making. Lastly, we saw the example of the Palestinian-Israeli peace talks getting derailed due to a lack of common ground in mapping. The next chapter will further explore the theme of maps and their messages, as we turn to what has reportedly become the most controversial map in cartographic history, the Peters.

Fig. 2-15A This is one application of the Camp David proposals. It illustrates the point made by Chomsky and by President Jimmy Carter and supported by many Israelis. The areas designated Palestinian territory would be effectively cut off from one another.

Source: © 2004 Washington Institute for Near East Policy. Reprinted by permission.

Map Reflecting Actual Proposal at Camp David

Legend:
- ▨ Proposed Palestinian State
- ☐ Israeli Settlement Blocs Annexed to Israel
- ▧ Israeli Security Border

MEDITERRANEAN SEA

Tel Aviv

Jenin

Tulkarm   WEST BANK

Nablus

Qalqilya

Ramallah

Jericho

Maale Adumim

Jerusalem

ISRAEL

Bethlehem

Hebron

Jordan River

Dead Sea

0   10
miles

Fig. 2-15B This map, according to Dennis Ross, reflects the actual proposal intended at Camp David. Clearly, our mental maps make a difference in how we make decisions. Surprising? Not really: maps always frame and shape the way we perceive reality. Their messages have power.

CHAPTER 3

Behind the Scenes with the Peters Map[1]

"Purely in terms of sheer numbers of distributed copies, Peters' world map may be the best known map in the world, excepting only the Mercator and possibly the Robinson."

Jeremy Crampton, in *The Cartographic Journal*

If the statement above is true – and we have no reason to doubt it – it's a remarkable story. First published in Germany in 1974, the Peters map broke into the English-speaking world in 1983 and is now widely available also in French, Spanish, Italian, Swedish, and Danish editions. United Nations agencies, religious communities, international development and humanitarian service organizations are prominent among major users. It continues to make its mark in schools and universities.

Not surprisingly, Arno Peters' work has encountered heated opposition as well as enthusiastic welcome. To develop a fully informed perspective on the map, it becomes important to know the man.

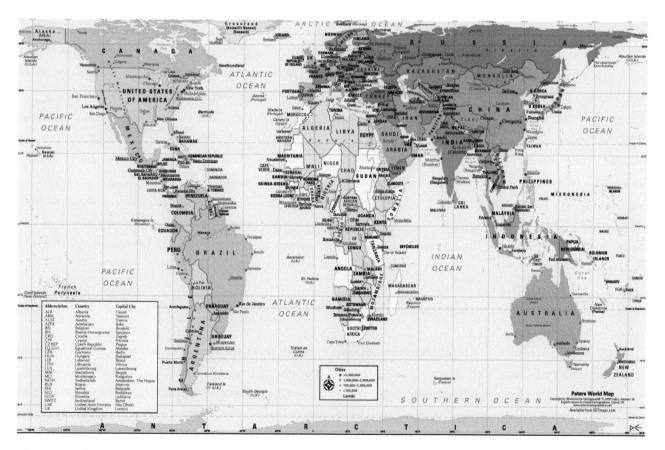

Fig. 3-1 Arno Peters' Equal Area Map. Originally meticulously drawn by a non-cartographer while he was working on his comprehensive world history project. The technical details were refined by Oxford Cartographers, who translated the map from German into English. See full size map on page 152

(From left to right) Lucy Peters, Arno, his brother Werner, Bruno Peters. Arno is about 5 years old here. He was exposed early in his childhood to social activists, as his parents, Lucy and Bruno, were highly involved in the labor union movement.

The Peters family had many visitors from other countries and other cultures in their home. Arno Peters' father, Bruno (missing from this picture), was imprisoned by the Nazi regime near the end of the World War II.

Shaping a Character

Born in Berlin, Arno Peters came of age during Germany's Nazi era. But other dynamics shaped him, notably his parents' example of commitment to social justice. They were independent thinkers and activists; indeed, Bruno Peters, Arno's father, was condemned to prison because he refused to conform to state totalitarianism. One of Arno Peters' early memories, he stated during an interview in October 2001, was of a black professor from the United States being entertained in their home (see also the Arno Peters Photo Album). The visitor, an author and activist in what we today might call the civil rights struggle or the liberation of the oppressed, left a lasting impression. This opening to a wider world, unusual in the Germany of that day, grew through other encounters that crossed the chasms of culture, race, and nationality. Together they tore at his loyalties. How could this 13-year-old German youth accept the official dogma of Aryan superiority? How could he go along with popular put-downs of other peoples? How could he deny his own experience?

As World War II drew to a close, Arno Peters saw firsthand the terrifying and tragic results of prevailing attitudes, of false world-views that led some people to deny the inherent dignity of others. Moral outrage flamed within him; he determined to do what he could to reshape the world in a fairer, more equitable way.

Righting Wrongs: The Historical Record

Even before founding the Institute for Universal History, Peters brought together an international, intercultural team of scholars to evaluate world history curricula. According to his analysis, these typically lacked balance and inclusiveness.

- They concentrated on Western history, paying only scant attention to the rest of the world.
- They weighted certain centuries heavily – mostly the recent ones – as if the accomplishments of other centuries didn't matter.
- They still used terms like "Dark Ages," seemingly unaware that the period they called dark was a time of great flowering of civilization beyond Europe's shores.
- "World" histories specialized in kings, political figures and battles to the virtual neglect of any deep understanding of cultural life or how the vast majority of ordinary people lived.

In short, the claim of mainstream historians to present a reliable view of the world's history was false.

Peters' bold response was to develop, over a period of years, and publish his *Synchronoptische Weltgeschichte* (soon adapted in a French version as *Histoire Mondiale Synchronoptique*)[2]. One might say that even as a historian he was thinking like a mapmaker, assigning historical data – events in time – a *spatial* value and *location*. Each decade, each century, from 3000 BCE to the present gets its own allotment of space on a page; there are no favorites based on arbitrary or personal preference.

This is a sample page from the monumental analysis of history, *Synchronoptische Weltgeschichte*. Here we see historical events and trends deemed most significant, year by year for hundred-year intervals. Readers are enabled to grasp *time* in *spatial* (not just in linear) terms, and to "synchronize" a variety of economic, cultural, religious, political, and military developments, allowing for integrated understanding. Source: www.heliheyn.de

Righting Wrongs: The World Image

But Peters was not satisfied. He contended that people's perceptions of the world suffered from geographic distortion as well as from historical skewing. And the Mercator projection, widely used, often uncritically assumed to be "the truth," had to be supplanted. Like the racist regime of the Nazis, it offended his sense of fairness to all peoples. Finding no satisfactory alternative, he set about to create one. It had to be an equal area map – absolutely no question there – as one might expect from a person of his convictions. He felt it had also to retain certain strengths of the Mercator, especially its rectangular grid, from which other desirable properties derived – including unambiguous orientation, north always being straight up and south always straight down.

Support and criticism soon surfaced. Dr. Peters reflected on the turmoil of that time:

> ...public discussion was such as had not been known in the history of cartography. I attribute this to the fact that the debate over my map was in reality not a struggle about a projection as such but about a world picture. Clearly, ideology had entered the struggle.[3]

And that points to what some consider Arno Peters' greatest contribution to the art and science of map making. In the storm of controversy that swirled around his map, that extended to his other work and finally to his person, it became clear that maps are never simply objective, scientific, mathematically precise, utterly reliable statements of "truth" – they are constructs, they carry a point of view, they have an agenda. To call that agenda political or cultural or ideological or commercial or practical does not alter the essential reality: the map has a purpose. It follows, then, as I have argued elsewhere,[4] that a map needs to be judged in light of its intent. Is the purpose worthy? How well does the map fulfill that purpose? To apply extraneous criteria is to run the risk of being irrelevant and unfair and, worst of all, to miss the point.

This is not to advance the claim that Peters was the only one – or even the first – to declare the need for an equal-area representation of the world. Rather, because of him the point got made in such a dramatic way that no one now can go back to what Peters called "the old cartography."[5] In a remarkable way, the world of map making has moved beyond the simple (though significant) split: pro-Peters and anti-Peters camps. Its taxonomy is now to be understood as pre-Peters and post-Peters. To borrow Jeremy Crampton's cogent phrase,[6] with the Peters Map "cartography's defining moment" had come. It is a matter of record.

Arno Peters lived, and the world is no longer the same.

For a perspective on Arno Peters – the man, his map, his creative message – we turn to Mexico. Diego Kloss, then professor of geography at the Universidad Nacional Autónoma de México (UNAM), situates Peters firmly among such leading thinkers as Galileo Galilei, Isaac Newton, Sigmund Freud, Karl Marx, and Noam Chomsky.[7] The point is not whether we approve the work each one did; rather, it is to acknowledge that the world is seen differently, and now works in a different way, because of them.

This book invites you, among its other objectives, to see Arno Peters and his map as changing the way we perceive and live in the world.

Arno Peters, age 64 (1980)
Source: www.ODTmaps.com

CHAPTER 4

The Map that Got Hijacked

"The Mercator Projection revolutionised navigation and has become the most common worldview."

Nicholas Crane in *Mercator: The Man Who Mapped the Planet*

A Low-Key yet Powerful Message

Pushing back a shock of greying hair, the old man leaned lovingly over the stamps spread out on his desk: British on one side, Canadian on the other. To the youngster beside him they held little appeal. Subtle pastel colors, monotonously similar in shape, all showing reigning monarchs who looked as if they never enjoyed a joke. One stamp, however, stood out: brighter colors, larger size, no king or queen to be seen. The old man lifted it carefully. "Here's a stamp I'd like you to have," came the offer.

Fig. 4-1 The Imperial Penny stamp
Source: Public Domain

The young boy sensed the *Wow!* in his grandfather's voice even before he heard the reason: "It has special meaning."

Just what made that one stamp out of a large collection so important?

To a philatelist it was doubly unique: besides being the world's first "Christmas stamp," it heralded a new era in global communications. Known as the "imperial penny" stamp, it enabled anyone, for the first time ever, to mail a letter anywhere in the British Empire for just a penny – all the way from Hong Kong to Johannesburg or from Gibraltar to Vancouver, for example.

But to my grandfather – yes, I was that youngster – the fascination of that little piece of colored paper lay at a deeper level. It featured a map of the world. *And what a world!* With Canada's vast expanse dominating the top central section and Australia anchoring the bottom left corner, plus the whole subcontinent then known as India (today's India, Pakistan, and Bangladesh) and chunks of Africa all flame-bright against the drab, receding grayness of the non-British world, this map made a powerful statement. "*We,*" Granddad seemed to be saying, "*we* are part of this great Empire!" That map defined us. It made us feel good about ourselves. We were big, powerful, important.

As you may have guessed, the map was on the Mercator projection.

The image was borrowed from a wall map published in 1893 by Bartholomew, a prominent British map house. The map itself was widely used in schools, especially in Great Britain and Canada. Titled *The British Empire Map of the World in Mercator's Projection*, it was the brainchild of George Parkin, a Nova Scotian whose burning ambition was to promote the Empire.

That map – and the 1898 stamp that gave it a big boost – must have set the hearts of many throughout the Empire beating faster. Granddad had lots of company in his chesty pride. "Look how much of the world we control!" may or may not have been actually said; it was surely felt. Britons were convinced they had a clear and noble calling: to take the benefits of civilization, of Christianity, of parliamentary democracy to the world. (The other side of the equation – the wealth we could extract from the far-flung empire – hardly got mention; it was simply assumed to be our right.) Studying the stamp through a magnifying glass, I could practically hear trumpets and massed choirs at full volume,

Rule, Britannia,
Britannia, rule the waves;
Britons never, never, never
will be slaves!

We were impressed with our own power. That power was conveyed not only in the might of the Imperial Army and Royal Navy; it came through an image that was uncritically taken as "truth." Nobody – so far as I know – ever cried "Foul! This is a map of mass distortion! It sets out a false view of the world and our place in it!" Who would dare call for time out, who would ask that the rules be changed in the middle of the game when your own side was winning?

Sending a letter half way around the world for a penny (or two cents Canadian) sounds, of course, like ancient history. Even licking a stamp and sending a letter – anywhere – seems slightly antiquated. The penny coin is today a collector's item. The British Empire is history. Mercator's world image, by contrast, is still with us. How has it fared?[1] What do people think of it? We turn first to cartographic professionals for their opinion.

What Professional Cartographers Say

From the same land that loomed so large on that stamp, the Royal Canadian Geographic Society says, "For about four centuries, mankind [sic] has been exposed to a distorted view of the world, largely due to Flemish-born German cartographer... Gerardus Mercator. The familiar Mercator projection planisphere gives a characteristically European-centred view of the world... that turns out to favour the north over the south."[2]

Arthur H. Robinson, sometimes called the dean of American cartographers, got right to the point. As an accurate picture of the world the Mercator projection, he said flatly, "is just terrible."

Such critical assessment was reinforced by the chair of the geography department at the University of Waterloo as far back as the late 1980s. "We have an extensive map collection," he pointed out. "But when I asked our librarians to pull for me all the Mercators, in response to your question, they had trouble finding any, apart from pages in books on the history of maps. No faculty member would ever use a Mer-

cator... well, unless they were teaching navigation or cartographic history."

"People's ideas of geography are not founded on actual facts but on Mercator's map," observed British cartographer G. J. Morrison in 1902, in words quoted several times since.

A high school teacher in Paramus, N.J. offered this observation: "I don't think anyone in our social studies department depends on the Mercator. There's so much wrong with it, I doubt any reputable publisher or teacher these days is willing to use it."

Rand McNally, a highly respected map company, declined to sell publishing rights to a Mercator image it no longer used. In their explanation of principle trumping profit, they said, "[The] North American centered Mercator projection conveys a worldview and corporate image that are contrary to [our] mission... We do not license or publish products that promote ethnocentrism."[3]

How Lively Can a Dead Map Be?

Some reviewers of my earlier book, *A New View of the World*, felt it gave too much attention to critiquing the Mercator. "It's dead," they seemed to be saying, at least as a general-purpose map.

"Nobody pays it much attention anymore."

My experience, frankly, differs. As an image of the world the Mercator is rejected by the experts – no question – but it remains strangely popular. If it's dead, it's as lively a corpse as ever was seen.

Among those who agree is H. Daniel Stillwell, professor emeritus of geography and planning, who in retirement took over a retail map store. He writes,

> My wife and I operate Blue Planet Map Company, a retail store that caters to university students and tourists. We found many cartographic materials in our store that use the Mercator projection: educational booklets for children, blank world maps, puzzles, inexpensive atlases, wall maps, outline maps, encyclopedia articles, and even scientific publications.... It turns up on shower curtains, wrapping paper, and other products.

Assuming that map producers have no particular agenda, I can only attribute the continued use of Mercator's projection in schools to ingrained tradition and simple ignorance.

As a former college professor, I am particularly concerned about the use of Mercator-projection maps by scientific groups. For instance, *This Dynamic Planet*, a large wall map issued in 1994 by the Office of Naval Research, the U.S. Geological Survey, and the Smithsonian Institution, is touted as "a teaching aid for classroom use and as a general reference for research." Its makers chose a Mercator projection "both for its familiarity and for the user's ease of transferring information from other Mercator maps" (thus indicating how widespread the use of this "perfectly awful" map is among scientists and educators).[4]

We'll return to the question of the Mercator's surprising longevity, but first we need to broaden our scope. The Mercator does not stand alone.

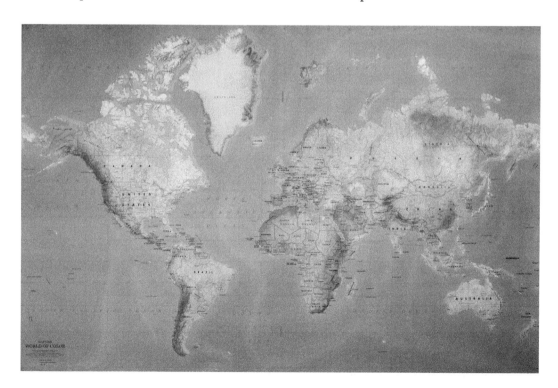

Fig. 4-2 The Mercator. Rapidly falling from favor as an appropriate way to view the world.
Source: www.ODTmaps.com

Mercator and its Look-Alikes: Down but Not Out

During a campus presentation a student asked how to quickly tell a Mercator world map from some others we had been looking at. Specifically mentioning Gall, Miller Cylindrical, and van der Grinten, she said they looked so similar. I answered briefly, then called on a professional mapmaker in the room for his own response. "There are clear differences," he began, "but it takes some real training and a sharp eye to see them. To the average viewer without benefit of cartographic expertise, the only way to be sure is to read the explanatory note that usually goes in one corner of the map. The name will be there."

He was right. Figures 4-2, 4-3, 4-4 illustrate the point that, in spite of their technical differences, some maps carrying different labels look remarkably alike. To facilitate discussion, let's call them the Mercator Look-Alikes, or simply "Look-Alikes"[5]... clearly identifying Mercator's map as the standard of comparison.

What I assert, then, is that, collectively, the Mercator and its "Look-Alikes" are, even today, a presence to be reckoned with.

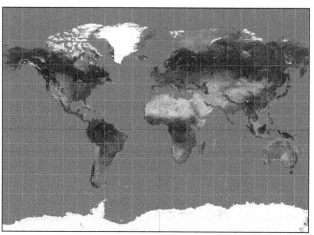

Fig. 4-3 The Miller Cylindrical Projection, developed in 1942. Practically identical to the Mercator at the Equator but with somewhat less size and shape distortion north and south of the Equator. The Miller Cylindrical normally includes Antarctica, which many maps on the Mercator projection will omit.

Source: Public Domain

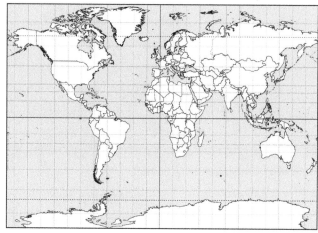

Fig. 4-4 The Gall Stereographic of 1855, often simply known as the Gall, also bears striking resemblance to the Mercator. It can be found in some British atlases, among other places.

Source: Len Guelke

Fig. 4-5a This projection, the van der Grinten (or van der Grinten I), by contrast, is not a "Look-Alike." It uses a curved rather than rectangular grid, and frames the result within a circle. It is clearly a different projection. Nevertheless, print versions (as shown in 4-5b) often cut off the polar regions and frame the truncated result in a rectangle, so reducing the distinction between it and the Mercator Look-Alikes. The National Geographic used this map for most of the 20th century.

Source: Public Domain

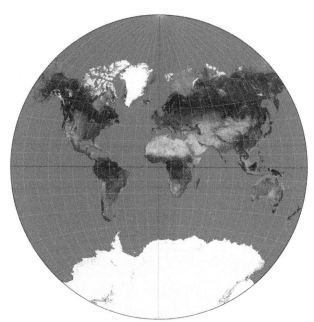

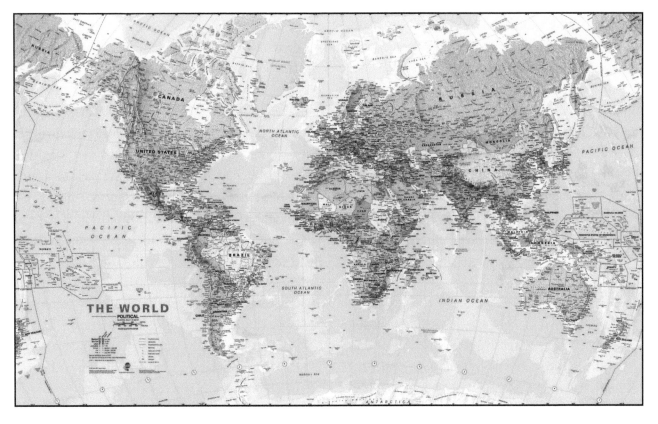

Fig. 4-5b This is also a van der Grinten Projection (like Fig. 4-5a) simply adding vivid colors and cropping the map to eliminate the curved grid on top, bottom, and the sides.

Source: © 2010 Lovell Johns Ltd

Comparing "Look-Alikes"

If your map shows parallels, i.e. lines of latitude (not all do), check the distance between them. The Mercator keeps increasing the distance between those east-west lines quite noticeably as the eye moves from the Equator toward either pole. While the Gall and the Miller Cylindrical also increase that dimension, both do it more gradually – actually by a precise but lower mathematical factor. As a result, they distort size less than does the Mercator. To gain this advantage they pay a price: they cannot show straight rhumb lines (lines of constant compass bearing) over the entire surface; hence they give up the Mercator's principal strength: its usefulness to navigators.

When used in atlases or educational settings, these alternative maps reduce (though do not eliminate) the Mercator's exaggerations. By showing a more familiar world, they clearly cushion the cognitive shock delivered by equal-area maps such as the Peters and Hobo-Dyer.

Whenever any of these "Look-Alikes" leave off grid lines, of course, the business of identifying the map becomes more problematic. That's when the average viewer must rely on information supplied by the publisher, or make an educated guess. Our rule of thumb for the "Look-Alike" category is to compare Greenland to Africa. In actuality, Greenland is 1/14th the size of Africa. If a given rectangular map shows Greenland nearly as large as the African continent, we call it a "Look-Alike."

Maps to be Reckoned With

The following list illustrates some of the ways look-alike maps intrude on our collective consciousness.

Books, Atlases, and Educational Settings

- PC Globe, a computerized atlas widely used in education and commercial settings, still – even in the 21st century! – relies on Mercator Look-Alikes.

- Webster's Online Dictionary uses a Mercator Look-Alike world map by permission of ClustrMaps, © 2009.

- The National Geographic, in its "Holiday 2008" and "Holiday 2009" catalogs distributed online and by mail, offers two laptop learning games, each based on a Mercator Look-Alike.

- Geocart, a Belgium-based publisher of maps and atlases serving customers worldwide, features a world map/Mercator Projection both for sale and as decoration in its catalog.

- Go to Google maps and – surprisingly to this observer – what pops up is a world map on Mercator's projection.

- High-Tech High, an innovative high school in San Diego, California, known for its demanding courses and innovative methods, still had a wall map in Mercator's familiar shape on its wall – as seen on the PBS *News Hour*, August 22, 2008. Part of the school's stated purpose is to close the achievement gap between America and its global competitors. In what way, one may ask, does a lopsided view of the world bring us closer to that goal?

- A Sunday School teacher in the American Midwest, ordering Peters maps from ODT in 2006, adds that the local school district "recently replaced all their wall maps throughout the system with Mercator maps!" Can you even *imagine* this in the 21st century?

- *Mandate*, published by the United Church of Canada, used Mercator images for years whenever they needed a world map. "These are the images most readily available from catalogs of stock photos," explained a spokesperson. And, one might add, the Mercator enjoys the advantage of instant recognition by readers. Nevertheless, that is now history; the magazine has decided to give the Peters map priority, states editor Rebekah Chevalier. Which means this story illustrates both the long-term hold the Mercator has on people – it remains popular precisely because it has been popular – and the possibility of shifting to a more appropriate equal area map.

- Widely respected naturalist Milos Radakovich, in a 2009 lecture titled "Fire from the Sea," uses geological history and plate tectonics to explain earthquakes, volcanoes, and the shaping of the world's land masses. When I asked why he chose the Mercator map to illustrate some of his points, his comment was revealing. "I have no particular love for that map," he replied, "I just feel it's important to let people see things through the perspective they're most familiar with." That's an opinion to be taken seriously, given his wide experience as a university professor and popular lecturer whose influence extends through radio shows and a CD that allows the Mercator to mingle with other maps.

- An Ivy League college developed a documentary to promote its mission; school officers used it with prospective students, donors, and alumni. The presentation was well received – until 2002 when one viewer pointed out that the Mercator image embedded was a poor choice for showing the world. The point is this: the product had been vetted by faculty and staff, it bore the prestigious name of the school, and it was seen by many groups that never called it into question. This strongly suggests that even among many trusted academics as well as representatives of the wider community, the Mercator image had become the accepted standard. There was no disconnect between the map before their eyes and the image in their minds. (Still, without diminishing the point just made, the story does have an important sequel: when the issue was brought to their attention, school officials acted quickly to replace the Mercator with other, more appropriate images. Credit to them!)

Mass Media

- The *News Hour*, a Public Broadcasting program, dealt with the world financial crisis October 15, 2008, and featured a world map that was unmistakably in the Mercator category. The same program, turning to the Olympic Games (Sept. 28, 2009), again used a Mercator Look-Alike.

- BBC World News, in a story on global warming September 30, 2004, illustrated the problem with an area-distorting map. Question: Isn't the *size* of area affected by temperature change more important than, say, shape of land masses?

- *Scientific American*, in an otherwise excellent article on world water supply (July 23, 2008), shows areas characterized as Abundant, Limited, Scarce, and Stressed supply. In such a visual, *area affected* would seem to be of primary importance. Yet even this esteemed magazine used a Mercator, a map whose strength is navigational reliability, not area accuracy.

- In a 2011 major exhibit on world water supply, the Royal Ontario Museum (Toronto) occasionally fell back on a Mercator Look-Alike, even while also using more appropriate maps.

- The German TV news program *Deutsche Welle* broadcast a report on how nations affect one another, using a Mercator map to show the interplay. This was in December, 2009.

- In the 2010 film, *The Book of Eli*, depicting a world in ruins three decades after a disastrous war, the wall map we see is a Mercator. Is it significant that the producers placed this map in the office of the villains? Or is the presence of this map a throwback, intended to suggest a world that once was, but is no more? Or does its use reflect mere carelessness or habit?

Commercial Applications

- CenturyLink, a communications provider, chooses to show a Mercator, which it then labels "The World Map" as if there were no other!

- MyCoolTools, a technical service provider specializing in internet and streaming video, until recently used a Mercator-based world map on its website.

- Publishers and other business leaders wishing to show world time zones regularly choose a map on the Mercator projection, since for this purpose a rectangular grid is highly useful (if you need proof of that, try showing the world's 24 principal time zones, plus variants, on a Robinson, say, or a Dymaxion World or a Goode's Homolosine). Since there are other maps with rectangular grids available, we respectfully raise this question: would it not be appropriate – even smart – to choose one of them to show time zones as vertical ribbons? Examples of the Mercator or one of its Look-Alikes being used for time-zone information are phone directories (Canada, U.S.A., U.K., and Germany for starters), Seiko's "Touch Sensor World Timer" clock, and Day-Timer's appointment books.

- MaRS, a Toronto-based think tank with global reach, still uses Look-Alike Maps – a striking anomaly that contrasts with its impressive work on the frontiers of science, technology, and social innovation.

- The Pelee Island Winery, located in Lake Erie, Ontario, flaunts a Mercator – some 8 feet x 12 in size – on its reception room wall. Thousands of visitors see it year after year yet no one, says a company spokesperson, has ever questioned it – suggesting that for great numbers of people, it is still seen as "normal."

- In a particularly ironic example, a "business summit" took place in June, 2011, with the aim of "strengthening investment and trade opportunities between India and Ontario." Fine – so far. But in their promotional material, planners flaunted a Mercator look-alike, so supersizing one partner – Ontario – and downsizing the presumably equal Asian partner. The contradiction between identifying India as "the world's fastest growing economy" – as well as the world's largest democracy – and shrinking it visually seems not to have hit home.

World Development and Service Agencies
In this category, let's be clear, there are fewer entries. This reflects the fact that by and large those who set policy for world development and service organizations are tuned in to values such as fairness and respect. Because they work for fairer treatment for all people they prefer a map that treats countries equally. Still, even some progressive agencies have yet to keep their maps consistent with their conscience.

- AmeriCares, a relief organization, prominently displays a Mercator on a brochure to show its global outreach.

- Heifer International, in spite of its focus on global relief and development, still uses maps that downplay the very areas they want to call attention to, a contradiction its staff has never been able to explain.

- Amnesty International USA occasionally reverts to a miniature Mercator-Look-Alike map in its publications. Interestingly, it "solves" what cartographers call "the Greenland problem" (its gross size distortion on Mercator maps) by omitting Greenland altogether.

Government Agencies
- Anyone applying for a Canadian passport gets a booklet of information. For example, the 2005- 2006 edition, "copyright Her Majesty the Queen," includes a spread of ten world maps, all on a modified Mercator.

- U.S. President George W. Bush had a Mercator Look-Alike behind him while he gave a National Security briefing on January 10, 2001.

- Regular background for press briefings in the early days of the second Iraq war was a world map clearly reminiscent of the Mercator.

- The United States Postal Service offers an information pamphlet on Global Shipping options, boldly and without apology using Mercator distortions.

- The G20 Summit of September 2009 in Pittsburgh created a huge map of the world as its "signature" symbol – on a Look-Alike projection. (For more on this, see Chapter 12).

A Personal Confession

Let's be clear about one thing: I do not criticize others' use of Mercator or similar images as though I stood above reproach. For years the Miller Cylindrical – one of the Look-Alikes – was my map of first choice. Curriculum materials I once prepared for Grades 9-10 featured a political map of the world on Mercator's projection for students to use.

Anyone looking for a way to understand that – or to excuse me – may point out that those were younger, more innocent days. Now with greater awareness, I hope I'll always choose the appropriate map for the task at hand, which means, in part, I'll never again take the path of least resistance, settling for a Mercator or one its proxies when I need a visual: there are better choices available! That's a promise!

What Arno Peters Learned

When Arno Peters faced this same question – How widely is the Mercator used? – in 1974, he uncovered some startling facts. Granted that the situation today has improved considerably, his comments are revealing:

> ...the central map office of the Federal German Republic (Geocenter) answered the question as to how far Mercator and Mercator derived maps control the map market with 'about 99 per cent'. In the great Readers Digest Atlas and in the small pocket atlas of Haack [a prominent German publisher] we find the old representation of the world unchanged... If one studies the map of the news on German television it shows the old map of the world... If one orders a global map stamp from the educational publishers Westermann for use in schools... one automatically receives the Mercator map;... and if the Federal German Post Office wishes to instruct [a customer] on air mail tariffs they offer a map of the world in the Mercator projection... and when foreign statesmen sign treaties... in the great assembly hall of the foreign office in Bonn, over their heads hangs the old map of the world from the colonial era...[6]

Analysts Provide Their Perspective

- In a professional paper prepared for the U.S. Geological Survey, John P. Snyder and Philip M. Voxland say of the Mercator, "Often and inappropriately used as a world map in atlases and for wall charts. It presents a misleading view of the world because of the excessive distortion of area."[7]

- Simon Winchester, author of *The Map That Changed the World*, says "Mercator's influence in creating a whole raft of political attitudes...is little short of astonishing. It is no exaggeration to say that Europe looked at his map, saw itself depicted so impressively, omitted to admit that it was a distortion, and promptly felt itself superior to most others on the planet... And America? North America, Mercator-blessed as well, has long looked and seemed so central to the planet's structure that to unquestioning consumers of Mercator's maps there is simply no questioning its supreme importance.[8]

- Matt T. Rosenberg, in a valuable online essay critiquing the Peters map, says the Mercator image "became the standard...in the mental map of most Westerners."[9]

Mercator Mentality/Imperial Mentality

"[O]ur understanding of the world is based, to a significant degree, on the work of map-makers of the age when Europe dominated and exploited the world." This is the assertion on the side panel of the Peters wall map. It is time to flesh out that statement.

Colonialism – substitute the term imperialism if you wish – is often identified with military conquest. As the Persians, the Mongols, the Turks, and others had done, Europeans dispatched their armed forces

– this time to Asia, Africa, the Americas, and the Middle East to take over major blocks of territory, resources, and people. But conquest alone is not colonialism/imperialism; it is only the entry point. A second, necessary part is economic: the human and natural resources of the conquered land are exploited. There is a vast outflow of goods and services; value is transferred from the weak to the powerful. A third phase is equally insidious: the work of colonialism is completed as the subject people internalize their subservience. When they see themselves as "people of a lesser breed," as small, weak, and having limited potential compared with the super-people who are their masters, then the work of imperialism is complete.

> **Either we give up our ideals of fairness or we get different maps. That is the choice before us.**

Even as some nations and people are diminished by size-distorting maps, some people in the supersized nations may be increasingly uncomfortable with what they see. Psychologists call this cognitive dissonance; it becomes intolerable to hold two conflicting views at the same time. You simply cannot reconcile the conviction that all people are essentially equal, and the unequal treatment they get on some maps. Either we give up our ideals of oneness and fairness, or we get different maps. The choice is that stark.

Any map that gives prominence to "colonizing powers" while minimizing "the colonized" in clear violation of actual size, may have other uses. It may be a good map for some specialized purpose. But it is totally unsuitable as an image on which to base relationships of mutual respect. And those who mindlessly continue to use it are in denial: they are living in a world that, if it ever existed, has long since passed into history. Their contact with the real world is tenuous at best.

The Hangover Effect

Sometimes people ask what would happen if we got rid of all Mercator Look-Alikes. It's a good question, as long as we understand it to mean as general purpose maps. No one – let me repeat – wants to do away with their legitimate use.

Would the world suddenly become a better place? Unfortunately, no. Like some binge drinker the morning after, vowing never to touch the stuff again, the world will experience the lingering effects of its long-term behavior.

And why and how do those lingering effects show up? A story may help. While I was living in France, colleagues told of seeing certain people stop on their daily errands, kneel, and cross themselves beside an old stone fence. Since there was no religious symbol in sight, and since at least some of the people were in any case only vaguely associated with a church, their behavior was puzzling. Then, one day, workers tearing down the wall uncovered the clue: there had been, at some earlier time, a shrine in that wall. It had long since disappeared under successive repairs but – and here is the point – there was a residual memory. The stimulus for the behavior was gone, but the response – even to an absent stimulus – had a life of its own.

I leave it to others better qualified to explain the phenomenon: conditioned reflex, culture lag or psychic imprinting or muscular memory perhaps. What I do assert is that Mercator-like images of the world will live on in people's minds long after the visuals have passed into history.

Maps of Mass Distortion?

If we were collectively invited to conjure up some sinister force in the universe scheming to undo our sense of unity with all humanity, that "devil" couldn't find a better visual aid than size-distorting maps. Why? Because such maps encourage the feeling that some areas are home to movers and shakers while others shelter mere pawns. In other words, this "evil power" could take what was intended to be – and in some respect remains – a major contribution (a useful navigational tool, supporting commercial success and saving lives), and transform it into a device for reinforcing existing biases: the feeling that some groups are innately superior, others inferior. The simple, originally innocent, distortion on the face of the map segues into a distorting perception of the world. You don't have to believe in a conspiracy theory to get the point: a projection system that perfectly served its purpose got hijacked. Size-distorting maps continue to be used for unintended purposes. Once this skewed perspective enters the psyche of both world leaders and average people, we become living instruments of mass distortion.

When things go wrong, there's always a price to pay. And that does happen, because the map as mindset is always more powerful, stubborn, and resistant to change than the map as document.

Is This a Problem?

Let's rephrase the question: How can it not be a problem?

Look at it this way. Life made me a man. It wasn't a choice I made, but it is my destiny. I fully, even gladly, accept it. But the moment my affirmation of male identity downgrades the contributions of women, I impoverish my life and totally distort the experience of being alive. Fate – or the Creator or the accidents of ancestry or the lottery we call the gene pool, call it what you like – gave me a Caucasian heritage. I accept that; indeed, cannot escape it. But the moment I minimize the importance of persons with a different history, I reject the richness of life.

Fortunately, mapmakers have given us many tools for understanding the world; in our time, particularly, they have helped us appreciate the importance of the *whole* world, not just the once-dominant North. The right maps help us deal with the world-as-it-is. This too is a choice we have.

Why then should it be a matter of indifference when large numbers of the world's people go through life supposing the real world resembles a Mercator projection map – or their own – lopsided view of it? Why should their understanding of the world be impoverished? People do count; their opinions count, their votes and choices count. Therefore their maps count. That was Arno Peters' conviction, and it's our thesis in this book.

CHAPTER 5

The Peters Paradox

"All truth passes through three stages. First, it is ridiculed. Second, it is violently opposed. Third, it is accepted as being self-evident."

Arthur Schopenhauer, German philosopher

In Chapter 3 we referred to the controversy that broke out after Arno Peters presented his map. Now it's time to flesh that out. What happened? Why? What was the charge, and what was the defense? Who was right? To what extent has the issue been settled? What does it mean for professional cartographers and for average map users?

First let's get the big picture. To achieve this let's look at summary statements from both supporters and detractors. These are a matter of record; we offer no further comment on them here.

What Critics Say

"a remarkable example of cartographic deception."

Derek Maling – *Kartographische Nachrichten* – 1974

"American Cartographers Vehemently Denounce German Historian's Map."

American Congress of Surveying and Mapping Bulletin – 1977

"The Peters map projection is a 'plague' on cartography."

John Loxton – *The Cartographic Journal* – 1985

"the Peters map projection is a 'provocative and mischievous' product of 'a thoroughly confused cartographer' whose work is 'perverse and wrong headed.'"

Philip Porter and Philip M. Voxland - *Focus* – 1986

"the Peters map is 'misleading' 'manipulative,' and 'falsifying.'"

German Cartographic Society – 1985

"the Peters map is a 'cleverly contrived, cunningly deceptive attack' against cartography."

Arthur Robinson, Prof. Emeritus of Cartography, Univ. of Wisconsin, 1985

"The land masses are somewhat reminiscent of wet, ragged, long winter underwear hung out to dry on the Arctic Circle."

Arthur Robinson – 1985

"in over 40 articles on the subject, cartographers have vigorously denounced a number of Peters' claims."

John Snyder (engineer retired from the U.S. Geological Survey) – 1988

"RESOLUTION: urging publishers to 'cease using rectangular world maps for general purposes.'"

American Cartographic Association – 1989

"Academic cartographers became both puzzled and enraged" at Dr. Peters' "preposterous assertions."

Mark Monmonier, Prof. of Geography, Syracuse University, 1996

What Enthusiasts Say

"I consider the Peters projection to be superior in several respects to all previous projections known to me and would support its use in schools and the media."

Prof. Carl Troll, former president, International Geographers Union

"My favorite. The coolest map in the world. I keep one right over my PC where I write."

Thomas Barnett, professor, U.S. Naval War College and author of *The Pentagon's New Map*

"The first honest map of the world."

Harpers Magazine

"In his world map Arno Peters combines all those qualities that are important for climatology. No other production achieves such a sum of qualities. Without a doubt, the Peters Map is superior."

Prof. Heinz Fortak, Director of the Institute of Theoretical Meteorology, Free Univ. of Berlin

"The most accurate map of the earth's surface yet drawn."

Oxfam America – world relief agency

"I never teach a class in comparative development without showing the Peters."

Prof. Kathleen Staudt, Chair, Dept. of Political Science, University of Texas at El Paso

"demonstrates more accurate and objective perceptions of the significance of nations in both hemispheres."

Elizabeth Judge, Exec. Dir., Texas State Board of Education

"a burst of brilliance that can be compared with any major breakthrough in any field of science... For the first time in history, almost, we are seeing on paper what our world really looks like."

Prof. Vernon Mulcansingh, chair, Dept. of Geography, Univ. of the West Indies

"the best education tool for showing us our place on Earth... coincides well with the teachings of the Bible and the church."

Arthur O.F. Bauer, Lutheran Church of America

"innovative" and "accurate"

The Brandt Report of the Independent Commission on International Development Issues

What Do You Think?

What you have just read is a small but fair sampling of what has been spoken and written in criticism and support of the Peters map. Now it's time to formulate a stance of your own – which need not, of course, be your final answer. On a scale of 0 – 10, with 10 representing your full approval and 0 your total rejection, rate each comment. On what basis did you reach your opinion? On which statements would you rather pass until you get more information? (Actually, any statement that has ever been made for or against the Peters can itself be critiqued. What we are dealing with, then, is a multi-layered complexity rather than a simple debate. Result? It's all right to develop your own opinion; you're allowed to change your mind; it's OK also to keep an open mind. The only thing that is not appropriate is to walk away from this chapter as if the debate doesn't matter.)

Critiquing the Criticisms

The principal objections to the Peters map may be grouped together. To that we now turn, this time adding some interpretation and comment.

The Peters looks funny. Robinson's comment about long winter underwear makes the point vividly, but others have felt something similar. I count myself among them: the first time I saw the map I turned down the offer to publish it. Who needed a funny looking map like that?

In reality, however, the point of making a map is not to remodel the world until we all agree that it looks "pretty" – whatever that might mean – but to set out some aspect of truth in a way that we can understand it, talk about it, deal with it. Note these words: not to represent *all truth* about the world, but whatever is determined to be most important. Just

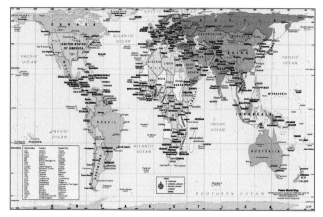

as Gerhard Kremer wanted to develop a projection that would make the seas safer for sailors – and succeeded – so Arno Peters sought to create a map that would treat all nations and all peoples equitably. In that he succeeded. The Peters, like any other map, is to be judged in light of its purpose. Since he did not set out to create a "pretty" map, to complain that the Peters (or the Mercator or the Fuller or any other) looks weird is to miss the point.

The Peters uses a rectangular grid. The planet is round, not rectangular. Therefore any attempt to show it in a rectangular frame is mistaken, runs the

reasoning of the American Cartographic Association in its 1989 advice to map publishers and users.

Point partially granted. Still, two comments must be added.

First, since the world is clearly more like a ball than either a rectangle or an oval, *anything* other than a circle – or, more precisely, an oblate spheroid - gives a false impression. This means that, to apply the full logic of the Cartographic Association's stance, most world maps in common use would have to be rejected. Even the National Geographic's Winkel Tripel and its predecessor, the Robinson, even Goode's Homolosine and Buckminster Fuller's Dymaxion. This was certainly not the Association's intent! The point is clear: an absolutist position such as this resolution takes may, in seeking to solve a perceived problem, get all tied up in its own contradictions. Rather, it seems to me that all maps – in all their dazzling variety – have a contribution to make to our ongoing conversation about our world – even if some people would prefer them to be housed in a different frame!

Next, even supporters of the resolution sometimes violate their stated principle. One example: The

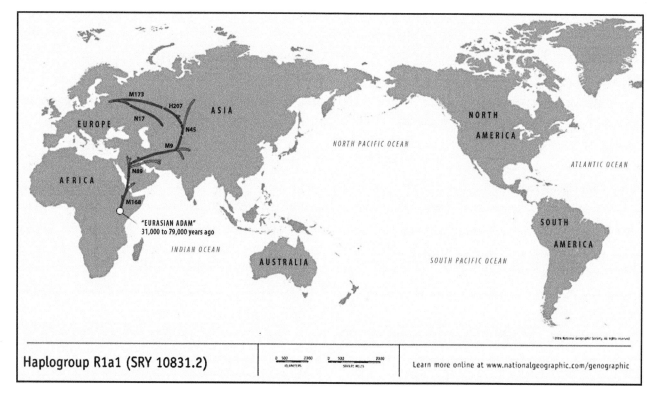

Haplogroup R1a1 (SRY 10831.2)

Learn more online at www.nationalgeographic.com/genographic

Fig. 5-1 This is how The Genographic Project, using DNA evidence, reconstructs how one person's early ancestors may have migrated.
Source: genographic.nationalgeographic.com

Genographic Project, of which the National Geographic Society is a sponsor, uses a world map in rectangular format and based on a rectangular grid (though grid lines are not shown) to illustrate human migration patterns. Similarly, on the National Geographic's website one finds world maps – in rectangular format! - used to show vegetation and land use and to situate smaller-area maps. Why would such reputable organizations violate their own advice? Is this a case of "do as we say, not as we do"? Or is the whole point about a rectangular format being wrong itself questionable?

The Peters promotes a cause. Shall we think about maps as disinterested statements of some objective truth that exists out there? Or do maps have a point to make or values to promote or an agenda to support? We have contended, with examples, in chapters 1 and 2 and elsewhere[1] that every map carries a point of view. Maps are inherently political. Some of their messages are obvious, others must be dug out; either way, there is a message.

The Peters is no exception: it invites us to think about the world and its peoples in terms of fairness and respect. Actually, it is quite up-front about that purpose; read the explanation on the right side of the popular wall-size version, for example.

In a free society anyone can object to a particular cause. People may, for example, find fault with the goal of fair play for all. (Some have gone so far as to label this map "communist" or "socialist" – therefore unworthy of serious consideration – precisely because it advocates fairness.) We begin with the recognition that all maps are selective in what they present, and that all maps have a purpose and a point of view. By what logic, therefore, can we take what all maps do, label that a fatal fault in the case of the Peters but a virtue in all other examples?

The Peters comes from a non-cartographer. That much is fact: Arno Peters did not carry a union card in the guild of professional cartographers. As we pointed out in Chapter 3, he was fully immersed in his landmark study of world history with its goal of filling in the empty gaps: the forgotten periods, the overlooked cultures, the neglected majorities of the people. When he needed a world map that would be as balanced geographically as his unique approach was historically, he found none. So he set about to create one.

One might well ask "What's wrong with that? Do we not honor those who find a need and fill it? Do we not recognize the contributions of gifted amateurs?" Bill Gates claims no M.B.A. behind his name, yet who would deny that he shows great skill in running a business? John P. Snyder was a chemical engineer by training; he never took a course in the ways maps can be projected, yet his impressive skills were front and center when the American Cartographic Association elected him to head their Committee on Map Projections. Vincent Van Gogh was a lay preacher without formal training in art, yet his paintings fetch record prices at auction. Maya Angelou came from a background of poverty where books were rare, but her literary contributions have been saluted in the White House and by those common people who, with a few dollars to spend, buy a book if it carries her name. Orville and Wilbur Wright worked in a bicycle repair shop; today we remember them not for the grease they got under their fingernails but because they obsessively pursued the possibility of human flight.

Arno Peters belongs in that pantheon. Do not assume, however, that creating a map was simple or easy. He worked diligently, even laboriously at his self-assigned task. He developed a number of variants before fixing on the one he introduced to the world in 1974. Peters was a perfectionist. He didn't start out as a cartographer, but he became one.

Some claims were unfounded. Some critics objected to certain of Peters' own claims; others, to statements about the map made by enthusiastic users.

An example of Peters' own claims is what he termed fidelity of axis. In this way he called attention to the fact that his map, like Mercator's and some other projections, was based on a rectangular grid. As a result, all North-South lines (meridians) intersect East-West lines (parallels) at right angles. This has an intuitive quality: it "feels right" to have a map laid out with the clarity of a compass. Such a grid, Peters contended, results in directional consistency across the face of the map. Some authorities have objected to that claim, contending that fidelity of axis is not a property of the map per se but simply a by-product of the projection system.

In critiquing the praise some have heaped on the Peters map, a number of authorities have pointed to what they call "overblown" or "meaningless" statements from enthusiasts. They point out – correctly, let me add – that *many* maps show the world "as it really is," therefore any claim that this map is a first, or the one and only of its type, or the best thing since somebody crafted the first wheel, is simply misguided. In this respect maps may be no different from other products or experiences – a play, a piece of music, a car – what people say about it may not be the final truth. We are free to listen – and to make up our own minds.

The Peters is a copy. The basis for this criticism lies in the striking similarity that the Peters map bears to one devised by James Gall, a Scottish clergyman, in 1855.

The question becomes: Did Peters copy what we know today as the Gall Orthographic or did he independently hit on the same solution? The two maps are virtually identical.

Investigators generally are convinced that Peters did not know of the existence of the earlier map. That is, he devised his map himself; he did not copy it. Seeking further clarification, some of us put the question directly to Dr. Peters in the fall of 2001. He gave an oral response, which I would summarize as an emphatic "No": he did not plagiarize Gall's work.

Some related realities are worth noting. For example, James Gall created his map, published it in an academic journal, then filed it away. Of his experience he says, " ... for more than twenty years after I had exhibited the three new projections before the British Association I was the only person that used

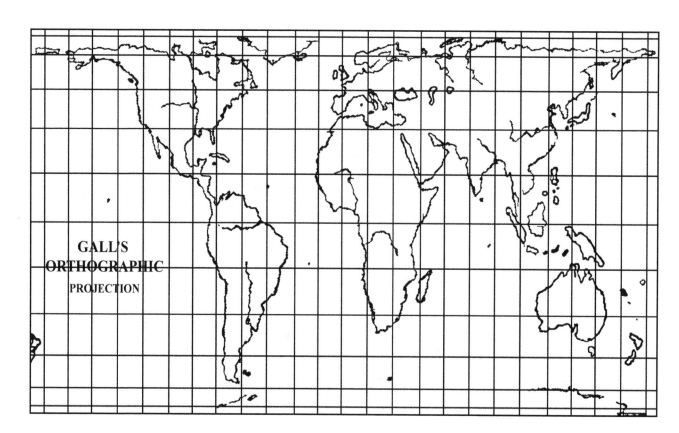

Fig. 5-2 Arno Peters created a map by hand that was nearly identical to Gall's mathematically-derived world map.

Source: Scottish Geographical Magazine, 1885.

them."[2] The U.S. Weather Service subsequently saw value in one of these – the Orthographic – and used it regularly. That was the exception, however. The Gall Orthographic became, we might say, the most important map nobody ever heard about.

We know also that Arno Peters experimented with several alternatives before hitting on this one. Specifically, where to place the standard parallels, that is, the lines of zero distortion. An early version of his map had these at 43° N. and S. before he finally set them at 45°. Though he tested a number of options, he decided to go with 45° – which means his map duplicates choices made by Gall.

Significantly, Peters committed himself to his dream of a better map, which Gall clearly had not done. Peters tied his map to a cause – the equal value of all peoples – in contrast to Gall's purpose: an academic exercise to achieve an equal-area map. Peters was not content to have constructed a map; it had to reach the people. As a result, where one map languished in obscurity, the other has leaped into prominence. The world that never heard of the Gall has learned to take the Peters seriously.

Peters was perceived as a threat. In lectures and in his writing, Professor Peters contrasted what he called "the old cartography" and "the new cartography," and in that analysis many mapmakers saw themselves being disparaged. Perhaps understandably, they did not take such comments lying down. As Mark Monmonier, himself a respected academic and cartographic expert, said, the criticism leveled by the cartographic community against Arno Peters and his map was "often irate, sometimes defensive, occasionally desperate."

J. B. Harley, a distinguished student of the history of mapmaking, analyzed the controversy around the Peters map as a struggle over power. "There is no doubt," he wrote, that "Peters' agenda was the empowerment of those nations of the world he felt had suffered an historic cartographic discrimination."

Was it also a question of who would wield mapmaking power: the establishment or an upstart? To the cartographic inner circle, to suggest that their work was anything but accurate, disinterested, and reliable was about as likely to make peace as waving a red rag before a bull. How could anyone – especially an outsider – dare talk about a political or ideological agenda lurking in the mapmakers' "values-free" results? If Peters prevailed, would not their prestige and authority be compromised?

> "I've got a saying: If all your peers understand what you've done, it's not creative.
> Henry Heimlich, M.D., after being criticized by other medical specialists for advocating the "useless" maneuver, now accepted worldwide, that bears his name.

Not only were Peters and his map attacked, sometimes even those who found value in his work were similarly marginalized. Thus Harley writes, "I was invited to publish a version of this paper ['Can There Be a Cartographic Ethics?'] in the ACSM [American Congress on Surveying and Mapping] *Bulletin*. After submission, I was informed by the editor that my remarks about the Peters projection were at variance with an official ACSM pronouncement on the subject and that it had been decided not to publish my essay!"[3]

Denis Wood provides his own take on the controversy. "It really is a shell game. When the aesthetic ['it looks funny'] issue gets hot, switch to science and talk about accuracy, but when that bluff is called, bring on the 'wet, ragged, long underwear.'"[4]

As Arno Peters saw things, it was a question of worldview. "... the debate over my map was in reality not a struggle about a projection as such but over a world view."

A struggle for power, comments by each side that the other side resented, a shifting battle scene resembling a shell game, a belief system under threat, a preference for the status quo over change, the need to defend the territorial integrity of an honored profession against the inroads of a maverick – you can find all these and more if you look for them. The final chapter in the story has yet to be written, but right now it appears that Peters has achieved some of his goals. His *bête noir*, the Mercator projection, is in decline as a world image. His map, on the other hand, is increasing in popularity.

A major moment in recent map history demonstrates just how far we've come since the days when controversy over the Peters was at its height. The Hobo-Dyer map (Fig. 5-3) is so similar to the Peters that most people have trouble quickly telling them apart without the clue of differences in color. It too is equal-area; it too accepts distorted shapes for the sake of achieving fidelity of area. I've never heard people call it "Peters II" or "Son of Peters," but such terms would make a point: it belongs in the family. It makes its contribution to the category of equal-area projections. Granted that fewer sweeping claims were made at its launching, the contrast remains strik-

ing: no brickbats have been thrown; there has been no public outrage; its creators and supporters have not been vilified. Mapmakers and map users seem to have settled back into a kinder, gentler mode; we may be moving into the third stage of dealing with new truth: acceptance.

If this means that a new understanding of the world is making its way into thinking and decision-making, the work of Arno Peters and his supporters will have been worthwhile. In that effort, even the contributions of Peters' critics are to be acknowledged, for without them the controversy that enabled his map to move to its place of prominence would not have been possible.

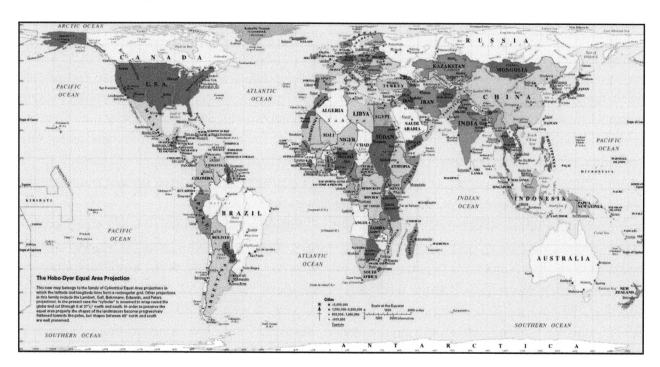

Fig. 5-3 Hobo-Dyer Africa-centered Map.

Source: www.ODTmaps.com

Full disclosure?

After I had signed the contract to publish the Peters Map in the English-speaking world, the news evidently reached Arthur Robinson. His simple advice: cancel the contract. The reason? The Peters was not a "good" map; it would soon be sent to the failed-maps graveyard. Neither my company nor I took his avuncular advice, as you may gather. Today I think of Professor Robinson as sincere in his action, but I wonder about his not mentioning that he also had a map to sell.

CHAPTER 6

A New Day for the World?

"The prime meridian, and the global system that derives from it, is commercialized, standardized, politicized – and enshrined in habit. But that is not to claim it is the best we could come up with. Read the story; is there a better approach?"

Ward L. Kaiser

Our planet doesn't come premarked with lines of latitude and longitude. In fact, neither gradations of distance nor units of time are a given. They had to be invented.

In the case of meridians – north-south lines of longitude – they were devised long ago. Running in a series of arcs from pole to pole, they divided the earth's surface into manageable units. Only one problem: there was no obvious starting point, no "zero" meridian. Or rather, there were many. The United States had one, running through the national observatory in Washington, D.C. France had one; it ran though Paris. Britain had its own, precisely located in Greenwich, a suburb of London. Spain had one, in Cadiz; Brazil's zero was in Rio de Janeiro. And so on.

As a result, ships at sea, using separate systems, couldn't signal position with confidence. Sailors not only spoke different languages, they calculated east-west distance from different zeroes. Words and numbers presented twin problems.

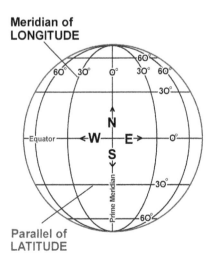

Fig. 6-1 Earth from space.
Source: NASA

The east-west lines mark latitude, and the north-south lines are meridians or longitude lines.
Source: Len Guelke

When Laissez-Faire Led to Chaos

The chaos of communicating location was overlaid with the chaos of telling time. Until well into the nineteenth century every community was free to set its own clocks. That hardly mattered as long as people lived out their lives in their own villages or valleys. Then, as rail travel grew more popular, people began to wonder about changing the system: could residents of different places not somehow agree what time it was? Could time be standardized? Even in a day when few people owned a watch or a clock, that seemed like a good idea. To get a sense of the problem brought on by imprecision, imagine a railroad timetable that gave arrivals and departures according to local clocks with their own varying times. Suppose, based on that timetable you expected to travel for two hours, your actual time between stations might be, well, 117 minutes – or 107 or 136.

The dream of getting everybody in a geographic area to conform to one official time found strong support in Sir Sandford Fleming (1827-1915), a Scots-Canadian engineer, explorer, surveyor, and mapmaker. Obsessed with the idea of standardizing time, he made speeches, wrote papers, cornered those he thought he could influence, lobbied and traveled widely to promote the cause. Eventually he was dubbed "the father of standard time" – deservedly so.

The Unity of Time and Distance

Here we need to set out the assertion that time – as measured on a clock or a calendar – and distance on an east-west line – are aspects of a single reality: the space-time continuum. To deal with one is to deal with the other. If that seems strange or surprising, think of two places at the same latitude; let's say they are 2000 miles/ 3200 km. apart. Obviously they have different longitude readings. Will they not also have different "sun" times: first light, high noon, sunset? Of course! This tight connection between eastwest distance and local time results from the earth's rotation. As the planet revolves eastward, places farther west experience their times later.

Could a way be devised to measure how much later, and to impose on the planet's unmarked surface a sequence of time zones that would be generally acceptable? Could Earth's space – that is, its geography or distance measurement – and Earth's time be coordinated in a single system?

To that end an international conference was called, set for Washington, D.C. in October, 1884. Its primary assignment was to fix the prime meridian: that starting point from which both distance and time could be measured. Twenty-six nations took part; in that simpler era that was the number of countries deemed self-governing. Fleming's adopted country, Canada, did not qualify, but Fleming was made an honorary member of the British delegation in recognition of his pioneering work. Before the doors closed on the three-week consultation, the Royal Observatory in Greenwich, England, had been chosen as the locus of the prime meridian.

That decision marks both a remarkable achievement and a deep disappointment. Let's take the upside first. The conference itself serves as a remarkable example of nations coming together to solve a problem that was spilling over national boundaries. Credit belongs not only to the big-name leaders of the event (including President Chester Arthur of the United States who, while not joining in the debate, did lend the conference the prestige of his office; and J. C. Adams of Cambridge University, renowned for having mathematically inferred the existence of the planet we now call Neptune though no one had ever observed it) credit goes also to the temper of the times. People in general were becoming more aware of their own identity, their geography, their history, their power, and of their relationship with other nations and cultures. Canadians took as their home a country just seventeen years old; in some respects they collectively resembled a teenager eager to

> **How long is a mile?**
>
> The difficulty of determining a ship's position at sea had its counterpart on land. For centuries the mile varied in length, based on local custom. Not until 1593 was it standardized, when an Act of the British Parliament fixed it at 5280 feet.

find out what their new freedom meant. The United States was experiencing unprecedented growth in population, industry, and philanthropy, and continuing its heady westward expansion. Europeans and Asians, no less than Americans, were thrilled at the prospect of submarine cables linking their three continents. There was growing support for negotiating rather than fighting, for international cooperation and peace. This was the era of "Give me your tired, your poor, your huddled masses yearning to breathe free..." – Emma Lazarus' 1883 poem that was later enshrined on the Statue of Liberty. It was an era of idealism confronting human need. The conference built on that. Holding the conference in Washington also signaled that the United States could be a player on the world stage.[1]

And delegates took a clear step forward. They imposed order on the existing hodgepodge of chronology and geography. By standardizing time the conference gave support to commerce and travel. It recognized the interdependence of peoples. Now there would be one system, widely accepted, for identifying location and talking about time; that agreement represented a major move toward international cooperation or what we now call globalization. Making just one change on the map – where the zero meridian would be placed – brought about a major shift in the way people dealt with the world.

Still, there was a downside. Some participants were far from pleased. Among the frustrated minority were Fleming and the entire French delegation. The French had made their position clear: the choice should be made on the basis of reason rather than political favoritism. "Let us place the crown on the brow of science," they pleaded, hoping to avoid unseemly competition among supposedly friendly nations. But disinterested science proved no match for commercial and political advantage.

As for Fleming, he arrived with a set of twenty proposals that he had presented and polished at earlier conferences in Toronto, Venice, and Berlin. This time he had impressive support from such astronomers as Otto Struve of Russia and trusted academics including Frederick Barnard, a geographer and president of Columbia University. His proposal was that "the prime meridian and time zero shall be established through the Pacific Ocean, entirely avoiding the land of any nationality."

Delegates voted him and his supporters down. Fleming was crushed. One member of the British contingent – a colleague of Fleming's – was particularly opposed. He zealously lobbied to set the zero marker in Greenwich, showing zero patience for any other idea. One hesitates to impugn another's motives – especially when he cannot explain them himself – but questions of narrow vision, national pride, and commercial advantage inevitably intrude.

Why, then, did a majority vote for Greenwich? Why not follow precedent and choose the Canary Islands, which had appeared as zero on some ancient maps because people supposed they marked the western edge of the world[2]? Why not the Great Pyramid of Giza (honoring an ancient civilization) or the Leaning Tower of Pisa (to honor the great Galileo) or Copenhagen or Paris or some other candidate?[3] Let's not overlook the commercial factor: British charts were much more widely used than any others. That meant two things: since British charts were based on the Royal Observatory in Greenwich, the prospect of moving the Prime (as the prime meridian came to be commonly known) anywhere else called forth strenuous objections not only from conference attendees but from chart makers. Their investment, after all, was at stake. And since non-British sailors had at least some familiarity with British charts, even if Greenwich was not their first choice, it was a less troubling alternative than some other candidates, so their opposition was muted.

Another factor was political. A full century after the split with Britain, Americans were collaborating with the British in ways that gave them a readiness to support Greenwich rather than some other contender. And Britain, then at the head of the most extensive empire the world had known, did call forth respect.

Choosing to locate the Prime in Greenwich meant that a second line had to be set. Regularly called the anti-Prime, it would serve as the International Date Line. Though some delegates pressed for the Prime and the Date Line to be the same, that was simply not possible in the case of Greenwich. It would never work to run the date line through a crowded, urban area! In reality, however, fixing the anti-Prime at the 180th meridian (counting either east or west) didn't solve the problem: it sliced through areas that logically should not be divided. Some Russians, some Americans, and some New Zealanders, for example, would be living in a different day from others in their own country.[4] To avoid this, planners gerrymandered the line – pushing westward here, eastward there as shown in Fig. 6-2. Indeed, the process goes on: in 2011 the people of Samoa voted to skew the line again so as to situate themselves on its other side. That is, precisely at midnight on December 29, 2011 they turned their calendars to December 31. Though they lost a day, they synchronized from then on with such major trading partners as Australia and New Zealand. Their action may be one more indicator that the decisions of the 1884 conference are not forever.

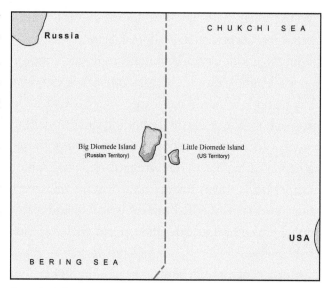

On a clear day you can see... the Diomedes – a group of islands in the Bering Strait – are split into two by the International Date Line. That means if you stand on Little Diomede, which is part of Alaska, you can look west about five miles, across the Date Line, to Great Diomede in Russian territory. In effect you are seeing "tomorrow." And Russians looking east can see "yesterday."

Source: flickr.com/photos/amapple. Credit: Andy Proehl

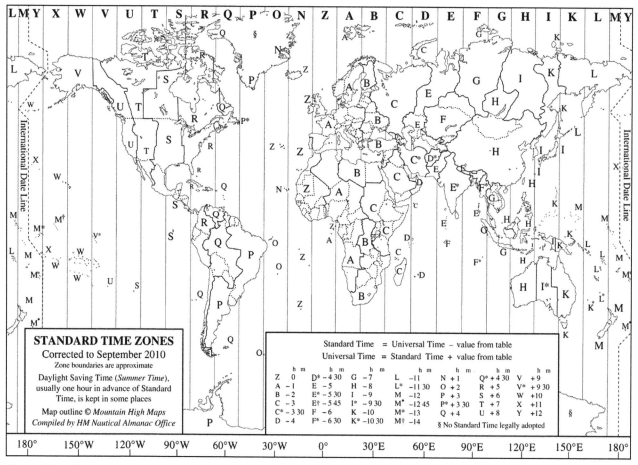

Fig. 6-2 Time Zone Map

Source: U.S. Naval Observatory (USNO)

If the conference proved imperfect in its principal focus, it may be said to have come up short in another respect as well. Not geographically, but in failing to set a new, high standard for international justice and respect. A mere month after the Washington conference, leaders of many of those same countries met to carve up Africa for themselves. The decision makers were mostly elite, elderly white Europeans, often without Africa experience. They had plenty of power – their decisions changed the future of a continent[5] – but little empathy. The people most affected, the Africans themselves, had no voice, no vote, not even a symbolic presence. The world still lives with the tragic results of their decisions; the boundaries we see on the map, the history of the continent, and daily news stories provide the evidence.

Fast forward ninety years. Enter Professor Peters.

A Bold New Proposal

A little-understood feature of the map bearing Peters' name is his proposal for a new prime meridian, a new International Date Line, and a new system of numbering both meridians and parallels of latitude. Let's consider these one at a time.

Try this for fun

Pick a location – your nation's capital, your school, your kitchen table, say – then develop a case for why the Prime should be located there. Should the International Date Line also be there? If not, where would it be – i.e., what is the meridian exactly halfway around the world from there? How workable would the result be?

Move the Prime

If choosing Greenwich was a political rather than a science-based decision in 1884, it may be even more questionable now. In point of fact, the Royal Observatory, originally marking the precise spot through which the Prime passed, is no longer even in Green-wich. (In 1948 it was moved to Herstmonceux Castle in Sussex and three decades later to Cambridge.) In any case, calculations no longer depend on measurements taken from any point on the ground; powerful computers, atomic clocks, and global information systems have taken over those functions. Obviously, the 1884 assumption – that the Prime had to pass through a major observatory – no longer holds. In principle, the prime meridian can be anywhere.

Straighten Out the Date Line

The only reason we have an international date line that is not the prime meridian, reasoned Peters, is that you can't split the week in an urban area like London. (Imagine the confusion if commuters left for work on Wednesday, arriving on Tuesday – things like that.) So planners transferred the problem to a relatively uninhabited area. But they still had to skew the line, as we pointed out, to make it work

But if we set the prime meridian in the Bering Strait, it can also serve as the international date marker. And it doesn't have to be manipulated. That is Peters' point.

*"Ladies and gentlemen, we just crossed the International Date Line. Have **another** nice day."*

Source: Punch

Adopt a Decimal Grid

Nearly the whole world uses the metric system of measurement, the United States being the one major holdout. That generalization, however, has to be qualified. The USA early on adopted the decimal system for its money: 10 mills = 1 cent, 10 cents = 1 dime, 10 dimes = 1 dollar.[6] And though milk and gasoline are sold by the gallon, wine and liquor dealers long ago gave up quarts for liters. Pharmacists measure medications in metric units. Every automotive mechanic depends on a set of metric wrenches, even for American cars.

Neither is the "metric world" consistent. Automotive and bicycle tires are sized in inches even in "metric" nations. Measuring the globe continues to be done in degrees (360 in total), which are then divided into minutes and seconds (60 minutes to a degree, 60 seconds to a minute). That system, using base 60 rather than base 10, was first devised by Sumerians some 4000 years ago, then spread by Babylonians.

Now look at the numbers along the borders of the Peters map. Note the one hundred "fields" of longitude marked across the top and bottom. Pick any number on the top border; take the same number on the bottom, and the line between them is a meridian.

Suppose someone asked you to find Nagoya, Japan, on a map. Even if you were situated continents apart, if you both had access to a Peters map your friend could say Nagoya is in longitude field 85. So find 85 on the top border, run your finger straight down and *voilà*! Or they could add the east-west field – in this case, 31 – to increase precision. In effect, the *process* is the same as in the traditional degree system (90 degrees north and 90 south, 180 degrees east and 180 west). The *numbering* changes, however, from hexagesimal to decimal. Each system is capable of fixing locations with equal precision.

Why Change, Anyway?

The 1884 Conference that set the prime meridian and the International Date Line also gave us the time zones into which the world is divided. Yet the idea of imposing order on the existing chaos – which today seems so obvious, so elementary – was a long time coming. Many influential people were apathetic or even resistant; they saw no compelling need to change things. And if things didn't *have* to be changed, why mess around?[7] Fleming's stubborn insistence on a coordinated, reasonable approach to clock time has clearly benefited the world. We all depend, whenever we travel to another country or another town, on the seemingly simple advance he advocated. Peters' proposals, on the other hand – to reset the International Date Line and the prime meridian, and to adopt a decimal grid for latitude and longitude – remain relatively unknown, even though his equal area map continues to gain acceptance. Both men's attempts to bring a reasoned order to their areas of concern thus connect in remarkable ways.

Is there a point there? We're at the earliest stage of this part of Peters' proposed new way of understanding and managing the world. Granted that the idea may be impractical or too costly or ahead of its time (or much too late), rather than cut off all discussion would it be prudent to look at it, air it in debate and assess whether there may be some value in it?[8] Did the French delegation at the Washington conference have a point after all: a decision based on rational considerations is likely to be superior to one based on other factors – even long tradition? Perhaps the next chapter in the story has yet to be written.

CHAPTER 7

Questions People Ask

When children call out from the back seat "Are we nearly there?" they prefigure more mature queries. "Where are we? If we're here, where is here? And what lies beyond the horizon?" The mind is a questioning machine.

Maps exist as a response. But they do more than give answers: The power of maps is their ceaseless ability to open up new questions.

Ward L. Kaiser

All maps stimulate questions.

At the most immediate level stand the factual questions: How far from this place to that? What bodies of water border Japan?

Some questions focus on function: how do maps operate? What influence do they have on us?

Others probe: why do maps elicit such a range of emotional and intellectual responses? Why do we consider some maps better than others? Exactly what does "better" mean? Since all maps provide information, how is it that some effectively expand our horizons? How do some maps challenge – or further entrench – our worldview or values?

This chapter contains material, largely verbatim, from radio and television interviews, lectures and classroom discussions, customer queries and personal conversations. Many of the questions specifically relate to the Peters projection world map, reflecting its demonstrated capacity to stimulate and provoke.

Q: Knowing about maps is a lot like knowing Swahili or algebra: it's nice, but who needs it? I mean, I can be a competent person – keep my lawn neat, pay my bills, show up for work – without stuff like that. So don't we need to be putting our energies somewhere else?

What I get from your comment is that nobody can do everything, so we need to set priorities. Within those priorities, where do we place the ability to understand maps?

Few complain that people know too much about maps. When we are told – as we were not long ago – that according to the polls, one-fifth of Americans cannot locate the United States on a world map, there's a lot more hand-wringing than rejoicing. Those statistics were quoted to Lauren Upton as she represented South Carolina in the Miss Teen USA contest in 2007. She commented, "I personally believe that U.S. Americans are unable to do so because... people... don't have maps."[1] Actually, you can watch her inept response, and judge for yourself.

In Canada, where geography holds a somewhat more prominent place in school curricula, the results are not much more encouraging: a 1988 Gallup Poll asserts that one in seven Canadians failed to locate their country on a world map.

Fortunately, the situation may not be as serious as those figures indicate. A National Geographic- Roper Public Affairs 2006 Geographic Literacy Study says that 94 percent of young Americans can point to the USA on a map. (How well they do on Iran or Tibet or the Nile River – well, that's another story.)

Whatever set of statistics you choose, the point is not just to feel good or gain competitive advantage for our educational systems. Two reasons strike me as more important.

First, we're talking about democracies – countries where the people take responsibility for voting in politicians, who support a particular policy. In the case of the USA, power intentionally derives, in the final analysis, from "We, the people..." and any elected leader forgets that at his or her peril. That includes international relations. So we all have a duty to be informed on the issues.

Furthermore, today we live in a connected environment: a world where what happens anywhere matters everywhere. Illness that strikes in an area we've never heard of can jet halfway around the world before we even know its name. Choices made in Ankara or Darfur, in Canberra or Beijing impact the lives of millions whether they can find them on a map or not. Similarly, vast numbers of people live or die based on business connections or foreign aid or emergency relief coming from, say, U. S. or Canadian or European centers of power. We live linked lives.

True, a person can get by in life without knowing much about maps. But let's ask, *Can they be good citizens? Are they full participants in creating a better world?* Karl Marx once said, "Our task is not [simply] to understand the world but to change it." I would ask, *Do we really think we can change it without understanding it? And how can we claim to understand it unless we understand what maps do to our perceptions?*

Q: **What are some ways we can overcome ignorance about geography and maps?**

Not knowing much about the world is without question a major problem. But I hold there is a more serious problem: misperceptions. Memorizing a mass of facts may be impressive, but it won't solve anything. *How* people *view* the world is crucial.

Specifically, do we see the world through the eyes of fairness and respect or through the eyes of privilege and power? This can show up in such practical decisions as what groceries to buy. Coffee is a prime example. A widespread misperception is that when we in the global North buy coffee we are helping people in the developing world: the more we drink, the more prosperous they become. The facts tell a different story, however, so "Fair Trade" enterprises such as Ten Thousand Villages, Dean's Beans, Pierce Brothers and – increasingly – retail stores and coffee shops like Starbucks seek to help consumers relate in a fairer way to those who do the work of cultivating, harvesting, roasting, hauling, and shipping. Are we willing to pay a little more for the sake of ensuring that farmers can provide food and a decent education for their families? Imagine for a moment the possibility of trading places with the coffee producers, or having your child trade places for a day with the child now harvesting the beans that go into your morning cup of coffee. Would that prospect – shifting from a distant, faceless connection to a more personal one – make a difference? Would your enjoyment of your morning cup of coffee be affected? The production of cocoa, the mining of diamonds, the weaving of carpets provide other examples of the system exploiting the weak and vulnerable, sometimes even fueling civil war. Our response to issues like this reflects our worldview, which in turn is shaped in part by the maps we use. To change the image is to take the first step toward changing the world.

"I work in the entertainment industry. I could tell you who reality show star Lauren Conrad's last four boyfriends were but I didn't have a clue where in Africa Maasailand is, let alone anything about it."

Jessi Cruickshank, co-host of an MTV show. Still, in an encouraging sequel, Jessi has become something of an activist for global humanitarian organizations such as Free the Children, "the world's largest network of youth helping youth." Her advice to young people everywhere: "We all need to stop thinking about me all the time and start thinking about *we*."

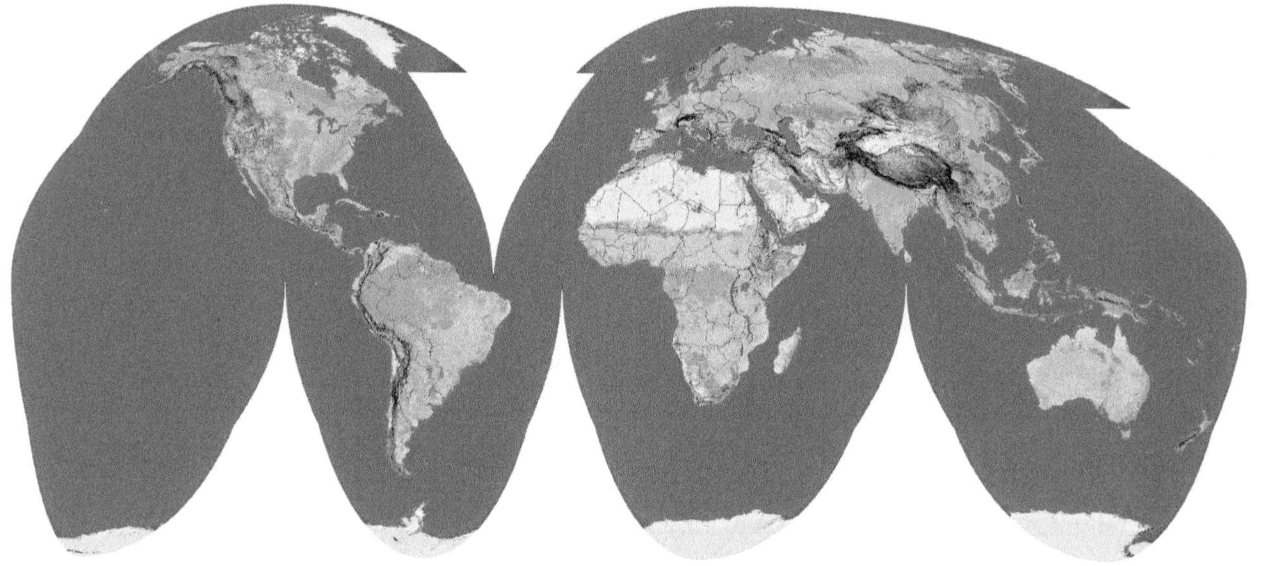

Fig. 7-1 Goode's Homolosine correctly shows land masses by size. See full size map on page 158.
Source: Oxford Cartographers

Q: One of the maps you have on exhibit for us is identified as Goode's Homolosine. Would you say more about it?

This is the map projection that is commonly described – sometimes with a smile – as looking like the skin of a peeled orange. That having been said, it is a map to be taken seriously. It was created in 1923 by J. Paul Goode of the University of Chicago. He developed it by combining two other map projections, the Homolographic and the Sinusoidal; by telescoping the two names he coined the term Homolosine. Goode's world map is "interrupted," that is, with empty space between segments or "lobes." The important thing to note about this map is that it is equal area. In other words, though you'd never mistake it for the Peters or the Hobo-Dyer, it shares that important characteristic with them.

Q: What is the thinking behind the choice of colors on the Peters map?

Colors are an important part of the language of maps. We did indicate that color may be used to provide political information, as when India was printed in red to show it belonged to Great Britain. Sometimes color can be based on political preference. Depending on the color chosen for the Malvinas/ Falkland Islands, or Tibet, or that part of Kashmir that is claimed by both Pakistan and India, or certain disputed areas in the Middle East, the mapmaker might well be making a statement of political preference. The Peters map deliberately sets aside colonial connections of the past in favor of present-day realities. One of those realities is the heightened sense of identity among the people of the world, particularly in Asia, Africa, and Latin America. Regional affiliations more and more take precedence over a relationship that owes its origin or continuation to forcible conquest and foreign domination.

A case in point: in the international concern over rigged elections in Zimbabwe (2008), it became critically important for other African nations to act. Zimbabwe is a former British colony, but history is not the same as present power. Britain's ability to pressure the Mugabe government was ineffectual, whereas neighboring African countries were in a stronger position to make an impact.

In his analysis of the world scene, Dr. Peters conceived the idea of showing a whole region in one dominant color-family, with each nation having its own variant. Thus the "family connections" as well as the separateness of each country can be shown. To my knowledge there is no other world map that uses color in precisely this way, or that takes regional awareness so seriously.

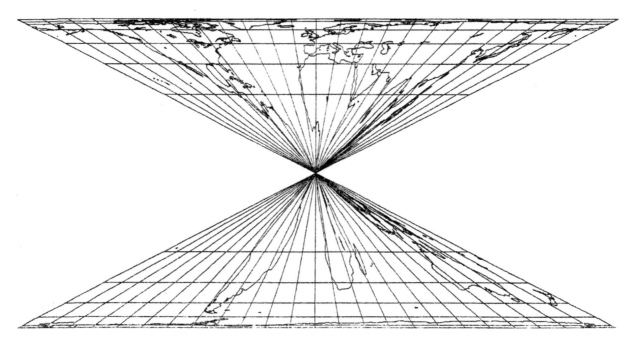

To Ward Kaiser,
This world map may be
equal-area, but it sure can't
geographically accurate (either)
Devised about 1945.
6/27/88   *John Snyder*

Fig. 7-2 Inverted Linear Equal-Area Projection

Goode, Peters, and the creators of the Hobo-Dyer were not alone in their efforts to achieve an equal-area image. This example – admittedly rather unusual – was created by John Snyder about 1945 and presented in 1988. Look carefully to recognize continents (Africa is set near the center of the upper triangle, for example). Shape is obviously compromised, but accuracy of area is maintained.

Source: John Snyder. From the collection of Ward L. Kaiser.

Fig. 7-3 Lambert Azimuthal Equal-Area (Centered on Chicago)

In 1772 J. H. Lambert created this equal-area projection in an effort to correct the Mercator's distortions. A mathematician and physicist by profession, he developed a number of maps that continue in use to this day.

Source: www.ODTmaps.com. © Len Guelke

Q: You and others claim a special place for the Peters map. Why not just see it as just one more map among all the others?

The Peters is one map among many – true. Still, it serves as what we may call a game-changer. With most maps, people can look, get the information they need, and move on. The Peters, on the other hand, regularly stimulates another level of response or engagement. Some viewers reject it and energetically denounce it. Others "love" it, advocating its widespread use. Not only does its appearance "shock," it has the ability to shift the way we look at the world. From that altered vision emerge new ways of relating to the world. It is this "social" dimension, this ability to open up new questions, that gives it an iconic quality, leading Prof. Jeremy Crampton to say that the launching of this map, plus the vigorous debate that followed, have given us "cartography's defining moment."

Q: Satellite imaging is becoming more and more important. How is that changing the way we deal with the world?

You're right in identifying satellite imaging as an important new tool in mapping and dealing with the world. It can lead to some unexpected results. Take, for example, the Hans Island affair.

Hans Island is a small outcropping of rock in the Kennedy Channel, between Ellesmere Island (Canada) and northern Greenland (Denmark). No one lives on it; no vegetation grows there. It has no known natural resources, and no tourists go there on vacation. Maps show it as part of Canada; the rest of the world couldn't care less – who would want it?

Along comes satellite imagery, and Danish authorities are convinced, they claim, that Hans is connected to Greenland. So they want it. They have even – several times now – planted the Danish flag there. Of course, the world's new awareness of the importance of the Arctic and its oil and gas resources plays into this. (Fig. 7-4)

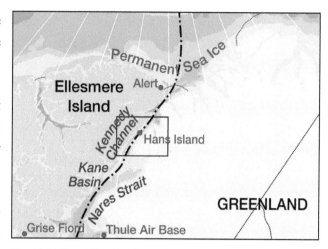

Fig. 7-4 Whose Hans? – that is the question as both Canada and Denmark compete for control.

Source: Steven Fick; *Canadian Geographic* January/February 2007.

The question of control has yet to be decided. One can hope that two peace-loving nations like Canada and Denmark will not resort to violence over a chunk of barren rock. In any case, killing or threatening to kill seems a pretty primitive response when maps are available – especially when they can claim the precision of satellite images.[2]

Q: It bothers me, what the Russians have done, planting their flag on the seabed under the North Pole. The Pole is part of Canada, so why doesn't the international community take action?

I'll answer your question with another: Doesn't the North Pole, in fact, belong to Santa Claus?

Actually, the North Pole is not in Canadian territory – it's at sea, ice-locked as that sea may be, beyond the territory of any nation. Your comment, however, points up a fairly common assumption among Canadians. Maybe it has been encouraged by Canada's national anthem with its soaring pride in "the true North, strong and free:" over time, people have sung the phrase and absorbed the idea that "true North" and "Canada" belong together. A map or a globe, however, tells a different story.

If, on the other hand, you ask who discovered the Pole – well, it certainly wasn't a Canadian. It was more likely an American – either Admiral Robert

Peary, who is regularly credited, or his African-American colleague Matthew Henson. Others hold that the Norwegian Roald Amundsen got there first.

Adding complexity is the fact that there is not one North Pole, but two: magnetic and geographic.[3] Magnetic north, the point to which a compass needle is drawn, shifts.

Whether at any given time it is situated in Canada or anywhere else carries no great importance. Geographic north, that one point on the earth's surface where you can look in only one direction, south, is what the fuss is all about. Of course, the issue is over more than bragging rights: it is clear that the Arctic region will play a major role in world affairs. Its oil resources if nothing else will see to that! The Northwest Passage – a shortcut between continents – is also lusted after. So the Arctic nations: Canada, Denmark, Russia, Norway, Sweden, and even the United States will each want to advance their interests.

Q: On road maps directions are easy to figure out: north is always straight up and east-west is 90 degrees from the north-south line. The size of a town or city is given, at least in general terms. In contrast, world maps can be frustrating. Sometimes north-south and east-west lines seem to come together at strange angles, so I feel as if I've got to tilt my head to figure out directions. As for size, if a map distorts that without offering any explanation or warning, isn't that like false advertising?

All world maps have limitations: that's our starting point. Most world maps do not claim to show accurate distances – except sometimes along the equator or some other chosen parallel – so for the great majority of users that's not a problem. If you want a world map where north-south and east-west relationships are as straightforward as on a road map, choose a map with a rectangular grid. That's an advantage enjoyed by the Mercator, the Miller Cylindrical, the Peters, and the Hobo-Dyer among others. Map projections such as the Robinson, the Winkel Tripel, and Goode's Homolosine use convex/concave lines for longitude,

so directional relationships are not as obvious or intuitive. In the opinion of some, the problem of distorting size is the most serious. Thus the authors of a standard geography textbook say about maps "relative area is of fundamental importance."[4]

This highlights your question about an apology: I believe publishers and promoters of maps that get sizes wrong should acknowledge that they distort people's perception. At the very least, they should warn consumers that certain maps may be dangerous to their view of the world.

Q: By the same logic, shouldn't those who support maps like the Peters make it clear that they don't show shape accurately?

Actually, in the case of the Peters that assertion has been forthcoming. In a handbook to the Peters map[5] I point out that it distorts shape, saying this is the price to be paid in order to achieve other qualities. Call it honesty if you will, or just stating the obvious.[6] Of course, you could say it's not on the map itself. But no map can do everything!

Q: The Peters would be more fully accepted if it didn't look so strange. Some even say it reminds them of Salvador Dalí's paintings: you know, watches melting over the side of a table and all that. Couldn't it have been made less different?

Can you help me know – perhaps with some examples – what you mean by the term "less different"?

Q: Take Africa... couldn't it have been depicted more normal – not so long and narrow?

Any mapmaker lives within certain constraints. First, there are the requirements imposed by geography itself. Any landmass must be shown accurately in terms of its coordinates, that is, those points where latitude and longitude intersect.

Peters chose a rectangular grid for a number of reasons, but partly because he believed most users were more comfortable with lines of latitude and longitude that consistently intersected at 90°. He also set out to achieve fidelity of area. Taken together, these goals impose strict requirements on the mapmaker. He or she cannot "fatten" Africa's midsection at will, nor reduce its top-to-bottom stretch without also adjusting other areas proportionally. A useful application of this principle may be found in the Hobo-Dyer world map. It does exactly what you hope for in how it presents Africa: it shows that continent less stretched north to south, or more "normal." The price it pays for this is to further distort, or "flatten", shapes in high latitudes.

Q: What are some ways people have used the Peters map?

There are some fascinating stories here.

The General Board of Global Ministries of The United Methodist Church had a Peters world map etched in glass at the entrance to its New York office. The result stands over six feet high; visitors cannot miss it as they come in. This serves as a fitting reminder of the agency's concern for the whole world, a world in which justice and equal rights will be enjoyed by all the people of this world.

The president of a major multinational corporation sent his chauffeur to the publisher's office, asking them to stay open after normal hours so he could get a copy in time for his next board meeting.

An interior decorator learned of the map through a design magazine; liking it, she recommended that it be hand-painted on the wall of a client's family room.

An organization devoted to international education uses the Peters map on T-shirts, promoting a realistic worldview.

When the Ontario Science Centre devoted its impressive space to investigating the question, **What Is Truth?** the Peters world map was featured prominently. (The accompanying photo shows a young woman pointing to the area of East Africa to which she had been appointed as an aid worker.)

Four management consulting firms – Wilson Learning Corporation, The Forum Corporation, KnowledgeWorkx, and ODT Inc., have used it in training sessions. "It frees up our participants," was one trainer's comment. "And whenever we divest ourselves of old ideas we open ourselves to new learning."

A teacher uses the Peters map to help students do two things: gain a more realistic view of the world they live in, and develop a positive, questioning attitude. "These days," comments the teacher from a Catholic high school in Toronto, "young people have so much thrown at them from all sides, unless they learn to sort it out and do some critical thinking for themselves, they're lost. Just because they see a Mercator on the wall doesn't mean it's the real world. I find we can take the discussion from there to other questions where they need to exercise some independent judgment. Just because some media idol says it's OK to do steroids or drugs doesn't make it harmless; just because the majority of their friends do something doesn't make it right. This map can lead to some amazing discussion."

Most people would have less dramatic stories. They would relate personal conversations about the map and the world it tells us about. They might mention putting it on a wall of their home, using it in college or post-graduate courses in geography or political science or ethics, or in informal settings such as world mission studies in churches. And not surprisingly Third World development action groups are major users.

"Who is civilized?" was part of the overall theme, "What Is Truth?" at the Ontario Science Centre.

Curators chose the Peters world map because it enables people to think in new ways about big questions.

This map has been prominently displayed in the Vatican and the offices of the World Council of Churches. NATO (North Atlantic Treaty Organization) forces have used it. UNDP, the United Nations arm that deals with development, supported the use of the map from its earliest days. UNICEF alone, which focuses on the world's children, has distributed over 60 million copies.

The map has been available in a variety of formats in ten or more languages. In addition to English and German, these have been Spanish, Arabic, Russian, Ukrainian, French, Italian, Basque and Asturian. In some cases interpretive materials are also available: in print, in audiovisual format or online. For example, UNICEF has published an atlas based on the Peters map for schools in Switzerland. An Italian company, in cooperation with UNICEF, developed a suitcase of print and audiovisual resources to support educational use of the map.

Q: Are Peters maps of individual countries available?

Friendship Press, the original publisher of the map in North America, published a map of Korea, for example. They also produced a map of the African continent. The *Peters World Atlas*[7] also shows a portion of the earth's surface – one sixtieth, to be precise – on any two-page spread.

At the same time, let's recognize that maps on other projections have met that need and will continue to do so. In other words, the value of the Peters is most evident in its whole-world version, not in maps of smaller areas. Useful and appropriate map projections of smaller regions already exist, so there is no need to duplicate them. Regional maps on other projections can combine area accuracy with correct shape.

Q: The claim is that the Peters map is fair to all peoples. But is that really true? What about, say, the Kurds in the Middle East or the Roma people in various countries of Europe – they aren't even shown!

Fairness – let's face it – is an ideal. We strive for it but it's always beyond our grasp. We cannot point to any nation, any political or educational or economic system, any business enterprise and say, "Look, that's what fair is all about!"

Given your interest in sports, let me turn to baseball for an illustration. Jackie Robinson, as you know, smashed through the color barrier. He was signed by the Brooklyn Dodgers in 1947, the first black player in the major leagues. Against great odds and at high personal risk he succeeded. Along with his manager, Branch Rickey, he pushed open the door for other blacks to play pro ball and to press for progress in other fields. Nine years later Rosa Parks sat on an Alabama bus, sixteen years later Martin Luther King, Jr. proclaimed to thousands at the March on Washington, "I have a dream" and 61 years later Barack Obama would attain the highest office in the land. The USA was striding toward justice.

Did Jackie Robinson bring about fairness for every victim of discrimination in the USA or the world? Of course not. But in words that came from his heart he summed up his view of how things work: "When you can't do everything, you do your part...and point the way for others."[8]

The Peters map, by giving Swedes and Mexicans, and the people of India and Iran and Colombia and Canada precisely their share of the world's landmass – no more, not one bit less – is as fair to all peoples as any world map can be today. We assert the same claim for the Hobo-Dyer and other equal-area projections.

Q: Making maps comes down to technical expertise, pure and simple. It's math-based, so either you get it right or you don't. That's why I can't go along with people like Peters or Fuller, who seem to be trying to change the way we look at the world. They have an agenda!

Let me affirm a point Denis Wood, Bob Abramms and I make in another book:[9] maps are never simple, straightforward representations of some abstract "truth." It has never been the case that "a map is a map is a map." What they leave out, what they choose to show and how they show it affect not only the visual product but the viewer's experience. Maps first frame our questions, then supply their own answers. Maps do take a point of view.

The sharp, sometimes acrimonious debate surrounding the Peters, previously unknown in cartographic circles, highlights the fact that *any* map carries cultural messages. In the case of the Peters, that comes through loud and clear. Brian Harley, an eminent authority on the history of mapmaking, sums it up: "[T]here is no doubt that Peters' agenda was the empowerment of those nations of the world he felt had suffered an historic geographic discrimination."

If we're honest about it, the debate has to do with our values far more than it does with "getting it right" in some mathematical sense. This is not to suggest that sloppy work at the technical level is acceptable. Rather, it is to focus on the underlying issue: do we support or reject – or are we simply indifferent to – those values that any map embodies?

Q: I don't understand why lines of longitude – meridians – are shown as straight lines on some maps but curved on others, though latitude lines show as straight. If the Earth is round, it's round in all directions, so shouldn't the lines that go around it all look alike?

A bit of history may help us here. A man named Gerhard Kremer wanted a map that would be useful to sailors. Their need was for a way to plot their course with reasonable accuracy. Such a map did not exist, so he set out to create one. To achieve his purpose he decided to show latitude and longitude with straight lines – only straight lines.

With that decision he divorced himself from the prevailing mode of mapmaking. Ptolemy of Alexandria, a cartographer of the second century, had established the ruling tradition. Though maps with a rectangular grid were not unknown, Ptolemy chose curved parallels, and curved meridians coming together at the poles. Then, in the Middle Ages Europeans, under strong religious influence, adopted a worldview centered on Jerusalem; when their cartographic creations reflected this they sacrificed some of the precision of earlier maps.

From a modern perspective, Kremer's dissent from prevailing patterns seems simple enough and totally logical, but at the time it was a daring breakthrough. He would lay the whole world out on a grid; every intersection on the grid would be right-angled. Using this projection, navigators could plot their course as straight lines and enjoy an enormous advantage and security they had not known before.

This rectangular format is still widely used. Kremer's creation, known as the Mercator Projection – or more specifically a version of it known as the "Transverse Mercator" – is the basis of navigational charts to this day. But when used in a whole-world presentation – never part of Kremer's intention – it has a major drawback: it distorts size more and more as you move from the equator to the poles.

Some mapmakers have therefore opted to return to a rounded grid. Result: size distortion is decreased. Examples include the van der Grinten and the Robinson (Figs. 7-5 and 7-6). These show one meridian as a straight line, termed the central meridian; all others curve inwards toward the poles.

What you say about lines of latitude being straight is partially true, but there are exceptions. Shall we say that mapmakers sometimes toss us a curve? The same van der Grinten from Fig. 7-5, is shown with the grid lines emphasized in Fig. 7-7. Note that though the formal name for lines of latitude is *parallels*, the lines on this map do not run parallel; they grow farther apart as they move out from the central meridian.

Now compare the two maps of North America (shown here as Fig. 7-8 and 7-9). That long stretch of the Canada-U.S. border that runs along the 49th parallel from the Great Lakes to near the Pacific may be shown as straight or curved, depending on the projection system. While either may be defended as accurate, in the curved example it is possible for people to get the impression that the border tilts northward as it moves west. Therefore in their minds Washington State pushes farther north than Minnesota, and Winnipeg, Manitoba, is located farther south than Vancouver, B.C. Compare locations for New York City and Los Angeles; can you understand how a casual observer might suppose they are at similar latitudes? The alternate map, Fig. 7-8, shows position more clearly.

In a day when schools offer so little geography education, the question is legitimate: Do we really want to add to such faulty perceptions?

Fig. 7-5 Van der Grinten projection

Source: © 2010 Lovell Johns Ltd

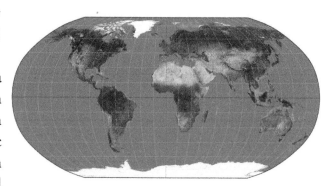

Fig. 7-6 Robinson projection. See full size map on page 154.

Source: Arthur H. Robinson

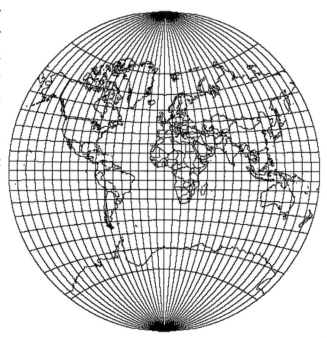

Fig. 7-7 Notice the grid lines on the Van der Grinten spreading farther apart as you move away from the equator.

Source: National Geospatial-Intelligence Agency

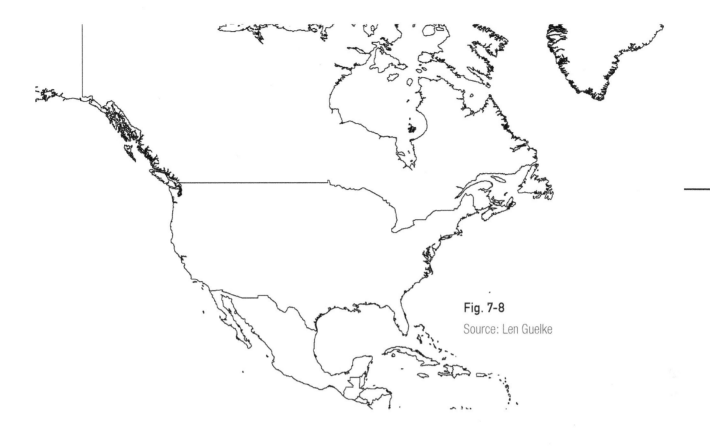

Fig. 7-8

Source: Len Guelke

Fig. 7-9

Source: © Len Guelke

Fig. 7-8 and 7-9 depict the same territory: the United States and part of Canada (though obviously one zooms out, adding the rest of Canada plus other areas north, south, and east). Same territory, yes, but how "different" they seem!

The second is more accurate for size, avoiding the severe distortion of Fig. 7-8 as the eye moves northward. But Fig. 7-9, by not showing lines of latitude, sacrifices precision in the viewer's perception of position, as the text points out.

Q: You've made it clear: there is no perfect map of the world; every map has some shortcomings. Since they're all flawed, what's the problem of sticking by the one we've always known ? [In the case of this Ohio high school, that was a traditional Mercator-like map.]

Isn't that like saying the perfect food doesn't exist, so we'll pig out on a daily diet of fries and Coke, thank you. Or, since by definition there is no perfect human being, I'll just hang out with whoever comes along... maybe we'll even get married some day. Either way, it's no big deal.

Common sense says there are differences between people; there's a difference between one kind of food and another. Maps also make a difference – a big difference! Which one is best depends on the purpose you have in mind. What do you want the map to do for you ? In your case, if your school has the goal of preparing students for a career as navigators, by all means choose nautical charts that conform to Mercator's projection. But even that, let me be clear, would not justify using a world map on the Mercator Projection in textbooks or on the classroom wall! On the other hand, if the school is supposed to be preparing students to live in the modern world, for heaven's sake junk your size-biased maps and go for something better suited to that need!

Q: I can't believe that Mercator-type maps distort people's sense of their place in the world. I'm a geographer; I grew up on the Mercator but I never thought Greenland was as big as it appears. Where is the scientific evidence?

My congratulations to you and to any others who may have managed, somehow, not to internalize the data and images that regularly confront us. I mean, if you consistently see things but don't let them influence you – well, that is remarkable, to say the least.

As for scientific evidence, do you always ask for that? How, for example, would you prove that "all men are created equal"? Yet since July 4, 1776 – when the Declaration of Independence was approved by the Second Continental Congress – Americans have affirmed it as true. What's more, some of us hold that the principle applies not just to men but to everyone, not just to Americans, but to people of all nationalities. So the Universal Declaration of Human Rights can affirm, in more sweeping terms, "All human beings are born free and equal in dignity and rights. They are endowed with reason and conscience and should act toward one another in a spirit of brotherhood."

Or turn to a more recent case. An anonymous graffiti artist sprayed the Berlin Wall with the words – in German, of course – "This wall will fall. Beliefs become reality." Skeptics might have insisted on proof, but that's not how some things work. Rather, the conviction over time provided its own proof: the wall did come down. Isn't the point that some convictions fit most neatly in the category of self-evidencing: they seem true, they are reasonable, they stand up to scrutiny, they generally meet the test of experience and they help move us toward important goals.

In my own experience of inviting people to draw a world map from memory, conducted among more than 200 students ranging from middle school through high school and college/university plus adults, in the United States and Canada, fully 80 percent produced maps that reflected size and shape distortions that owe their origin to the misuse of Mercator's work. Geoffrey Underwood, speaking for the National Geographic Society, similarly states that children asked to draw a world map regularly show an oversized Northern Hemisphere. Asked whether National Geographic world maps contribute to this distorted view, he said yes – "quite a bit." My statistics and Underwood's general statement are compatible with those of Thomas Saarinen's more formal research. Analyzing results from 75 universities in 52 countries, Saarinen noted that our mental maps generally exaggerate the northern hemisphere and diminish areas near or south of the equator.[10] Saarinen's work may be the most objective evidence now available to show that traditional maps do strongly influence the way we see the world. Therefore anyone who holds that the images we are inundated with are powerless to mold our mindset and shape our action – that person is making an assertion that really cries out for proof![11]

Q: There's an economic crisis going on right now; people are losing their jobs and their pensions –and you want us to think about *maps*?

Just as the present crisis affects not only decision makers in banking, insurance, and the housing industry, but clerks and carpenters and students, and farmers from Minnesota to Madagascar, so it will not be solved by financial wizards and politicians alone. The interconnectedness of all peoples is the basic reality here. Everyone – seven billion of us – deserve hope, the opportunity to better our lot, and fairness – in any stage of the economic cycle.

How do we help people see that better world? Without being dogmatic about it, I would find it hard to think of any major shift in world affairs in which maps did not play a significant role. So, yes, I invite you to think about maps, even when times are tough!

Q: It's OK to be idealistic, but when I'm faced with foreclosure I'm not going to care about other people's problems. I'll be totally focused on my own needs.

Nobody is required to put their own life – or that of their loved ones – in jeopardy to help others. Self-preservation is part of who we are.

Still, people can put service above self-interest. The news out of Haiti after the disastrous earthquake of 2010 included a story of a search-and-rescue team. It happened to be from Iceland. Now the remarkable thing is not that they traveled thousands of miles from a land that fits tight against the Arctic Circle to a tropical country, but that the sending nation was

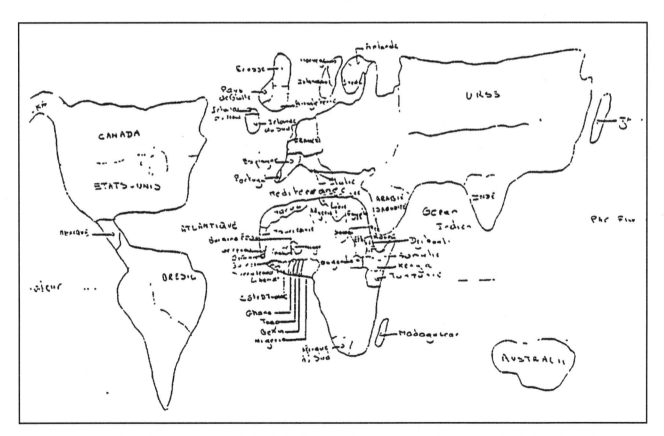

Fig. 7-10 This map, drawn from memory by a student in Thailand, clearly exaggerates northern nations. It is typical of mental maps Thomas Saarinen found even in Third World countries. He concludes: "A colonial mentality and Euro-centric image of the world still remains dominant in many places a quarter of a century after the end of the colonial era."

Source: www.eric.ed.gov

tottering on the brink of bankruptcy. They could have decided they had lots of work to do right at home. Instead, they recognized the plight of fellow human beings, and responded.

Many of us can probably cite examples of some pretty needy people who don't wait for the wealthy to carry the humanitarian load alone. In fact, Jay Brandenberger, associate professor of psychology at the University of Notre Dame, points out that as a group the poor give a higher proportion of their income to charity than the middle class.

Beyond the stories and the impressive statistics lies what I believe to be demonstrable fact: people crave community. Both Robinson Crusoe and Friday lived fuller lives after they found each other. That need for connectedness is as evident among nations as it is in our personal lives. Facing foreclosure can be terribly threatening; what can help is for people to experience the support of both friends and experts in finance. Being consigned to live at the bottom of the world's disastrous economic system, with virtually no hope of climbing out, is – if we can imagine it – frightening and frustrating in the extreme. Only if people begin to see the world as a community will we progress toward solutions.

Q: The most accurate portrayal of our world is one we already have. It's called a globe. Why don't we all just use globes and forget about world maps?

You are right to point out that a globe, and only a globe, can be fully accurate. Both shape and size can be correctly discerned on it, for example. Even a cheap, inflatable globe can be a helpful adjunct to a flat map.

Nevertheless, any globe has severe built-in limitations. For one, how would a teacher use a globe in class? The average globe is too small for more than two or three students to use at a time. To enable a whole class to examine a globe at the same time one would need to build a room around it, since so large an object would never fit through an ordinary doorway. Even then, students would see less than half the world at any one time. Its equal-area advantage would be lost every time students could not see, at the same time and on an equal plane, the several places to be compared. Globes have a portability problem as well. So the list of inherent deficiencies continues.

Still, there is hope. John Hammer, former president of the International Map Trade Association, suggests that a screen with a projected globe on it might suit admirably. (He's right!). In the future new holographic technologies may come up with solutions to the "flat map vs. hard-to-display globe" dilemma. But as far as we know, we're not there yet.

Flat maps are needed – no question. Many maps help meet our human hunger to understand our world with precision and clarity.

Q: You present an interesting case, and I have learned from it. My criticism is that it lacks balance. You have a point of view and you make it clear, but wouldn't it be better to keep that in the background... present only facts and let your readers make up their own minds?

What I am attempting here is to show that maps have a social component, a political component, an ethical component, a personal-perspective component. My assertion is that the scientific-factual base should always be there, but the point is to go beyond that to explore the implications of our maps. And I hope that in allowing my own convictions to show through I am following impressive precedent ...After all, where would the world be without the courageous stance of Nelson Mandela or Martin Luther King, Jr.? Or Rigoberta Menchu, the aboriginal peasant woman who as a child labored in the coffee fields of Guatemala, whose story of courage confronting power won for her not only the Nobel Peace Prize but the assurance that she was helping to right wrongs on behalf of countless other "forgotten" people? Where would we be without the example of those Hebrew prophets who dared disturb the status quo? Advocating is not a bad word! Still, people are free to make up their own minds. I hope you will!

CHAPTER 8

The World We Want

"Over here will be water...and this part is a place for elephants to get a drink." "This is where the people live, so we have to have lots of houses." "I'm building a hospital, so sick people can come and get better..."

from an unrehearsed conversation of a group of children at play, setting up their own mini-world.

Children – that's who we are. We play with ideas; we imagine a different world. Science fiction writers spin out such dreams on paper or film; prophets, activists, and mystics of many backgrounds broadcast their beckoning visions through spoken words and sacred writings. Academics simulate alternate futures, stimulating debate and action. Video game aficionados enter a different, virtual reality. And in the windup to every election, voters ponder political platforms, asking the essential question, What if ...?

Inventing the future is serious business. It's also play. It's the kind of game where, if we get it right, everybody wins. What's not to love about that?

Now it's time to ask, "What kind of world do we want, anyway?" Assuming it won't be a clone of the world we have, just how will it be different?

This chapter takes a different approach from those that have gone before. It is more focused on process, allowing your input. If you're using this material in a group setting, you'll reap the benefit of others' ideas. If you are reading it by yourself, the steps outlined here can be useful as we continue our "conversation" about maps and what they mean.

What's Your Vision?

Stage One: Right Brain Time / Brainstorming

I urge you to take a few minutes right now to pull together your thoughts on the kind of world you'd like to have – or see or live in. Keep in mind that the invitation is not to imagine a better job or even a better town but to think globally – the World We Want. So get a pad of paper and a pen. Describe your vision in a poem or a paragraph, or song, or draw a picture to sum it up. Liberate your imagination – let this be "play time." Don't get bogged down with details or practical questions like how to pay for your dream or even how you'd get people to vote for it; all that can wait. If you're part of a group, share your statistics or artwork or catalog of characteristics – whatever you might come up with. What follows is geared to a small group; if you're reading this on your own you can (let me repeat for emphasis) still follow the process profitably.

Stage Two: Engage

Questions for clarification are in order at this point; critiquing or being judgmental are not. If someone's concept does not embrace every aspect of life, consider why what they *did* present is important. Thus you might ask, "Why do you think the five big ideas you've listed are crucial?" But bite your tongue before blurting out, "that's the stupidest statement I've heard all day!"

Stage Three: Find Patterns

Clusters of ideas will emerge. Whether under the guidance of a designated leader, by democratically sharing insights or exercising your own intelligence, find those patterns. How many ideas deal with economics, for example? Politics and forms of government? If several contributions focus on how children are cared for, consider putting them together under the same umbrella. In this process some informal evaluation is certain to take place. What ideas get you or your group excited?

Do some ideas seem crazy or impractical? If so, you're in good company. Mark Twain once proposed a way to solve all traffic problems: no car would be allowed on the road unless it was paid for. Camelot may turn into a vision of a world where candy has no calories, horses are born with saddles, fish jump out of the water, and rain falls only after sundown. Isaiah talked of a peaceful world in which "a little child shall lead them." Can anything be more far-fetched?

While holding to the ideas you've come up with, consider the following vision.

Bono on the World We Want

Every year in Washington, D.C., there is an event known as the National Prayer Breakfast. Bono, the popular Irish rock star, was invited to speak at the 2006 event.

Bono applauded Americans, through the elected officials seated before him, for their good work.

> [Y]ou have doubled aid to Africa. You have tripled funding for global health. Mr. President, your emergency plan for AIDS relief and support for the Global Fund – you and Congress – have put 700,000 people onto life-saving anti-retroviral drugs and provided 8 million bed nets to protect children from malaria.

But Bono boldly went on.

"Here's the bad news," he said.

"Equality is a real pain."
Bono

It's not about charity, is it? It's about justice.

Let me repeat that: it's not about charity, it's about justice.

And that's too bad. Because you're good at charity. Americans, like the Irish, are good at it. We like to give a lot, even those who can't afford it.

But justice is a higher standard. Africa makes a fool of our idea of justice; it makes a farce of our idea of equality. It mocks our pieties, it doubts our concern, it questions our commitment...

Because there's no way we can look at what's happening in Africa and, if we're honest, conclude that deep down, we really accept that Africans are equal to us. Anywhere else in the world, we wouldn't accept it. Look at what happened in South East Asia with the Tsunami. 150,000 lives lost... In Africa, 150,000 lives are lost every month. A tsunami every month. And it's a completely avoidable catastrophe.

It's annoying but justice and equality are mates. Aren't they? Justice always wants to hang out with equality. And equality is a real pain.

You know, think of those Jewish sheep-herders going to meet the Pharaoh, mud on their shoes, and the Pharaoh says, 'Equal? A preposterous idea: rich and poor are equal?' And they say, 'Yeah, "equal," that's what it says here in this book. We're all made in the image of God.'

And eventually the Pharaoh says, 'OK, I can accept that. I can accept the Jews – but not the blacks.

Not the women. Not the gays. Not the Irish. No way, man.

So on we go with our journey of equality. On we go with our pursuit of justice.[1]

Now let's analyze what you've just read. How would you describe the world Bono wants? What words come to mind? Would they include – for starters – hope, truth-telling, one-sided, selective, brainless, politically correct ["P.C."], a fair chance for all, tough love, respect for all people?

Where does Bono's vision differ from the one you set out? For example, do you hold that taking care of the natural environment should be a high priority? (Bono didn't mention it; is that a problem?) What about energy independence? Does your ideal world call for a Hummer – or at least a hybrid – beside every hut? If not, why not? How do you feel about the direction some emerging economies are taking: more cars, more global competition for oil, more American-style consumerism, greater independence in setting policy? What does your position on these questions say about you, about your commitment to saving the planet, about practical considerations, about equality? Should there be more – or fewer – private vehicles? If you considered education, should it be free? Available to all? Should it include basic literacy? How to manage money? Values education? Critical thinking skills? What's the best way to finance it? What did you say about health care? War and peace? Race? Gender discrimination? Financial security? Freedom from hunger? If security got onto your list, do you see it in terms of an alarm system in every home, a guard dog at every door and a pistol under every pillow, or a healthy investment portfolio, or good relationships with neighbors or fewer guns or a world of peace and respect? What did you say about water: do *people* have a right to safe drinking water, or *fish* to unpolluted streams? What role do you see for international agencies such as the United Nations, the World Court, The United Nations Children's Fund (UNICEF), the international police clearing house known as Interpol, the World Bank, the body set up to deal with refugees (United Nations High Commission on Refugees or UNHCR)? Did your vision mention borders, trade and tariffs, terrorism? How about the issue of which nations should be allowed to have and hold nuclear weapons, and which ones should be kept outside the nuclear club? (In this instance do you vote for equal opportunity for all? Or no nations at all? Or only peace-loving nations in the nuclear club? Who should decide?) To put the question another way, if you had opportunity to follow up Bono's speech, what would you say?

"A World in Space"

Bono is not alone in envisioning a different world. As we pointed out in the opening lines of the chapter, imagining a better reality is part of who we are as human beings. Here's a provocative statement attributed to Rabbi David Polish, prominent leader in Reform Judaism:

There is a world out in space which is an exact duplicate of our own. It is populated with men and women like ourselves. They live in countries like our own. They live under various economies and governments, and are divided into different national, religious, and racial groups. They differ from us in only one respect. In each country there is a pathological obsession with human welfare.

As a result, over 60 percent of the national budgets are devoted to a compulsive and hysterical desire for sheltering life from the normal ravages of human existence which we accept more stoically. Billions of dollars are spent by governments on the conquest of disease. Over the years, nations have poured their resources into medical research and today no cancer, no kidney ailments, no degenerative disease exists.

Unheard of sums are spent by governments on housing. They have so ordered their fiscal policies that slums and blight are unknown. They are so overprotective of their children that they overpay teachers, and training schools for teachers have to turn candidates away. The perverseness of these conditions reaches its greatest height in their legislation against all private charities in behalf of human welfare. The outlawing of private charity has, of course, stifled the philanthropic instincts of the people.

There is only one exception to this restriction against private benevolence. Since the national budgets are so swollen with human betterment appropriations, there is little left for national defense. It therefore becomes necessary for private citizens to raise money for armaments. Thousands of private organizations exist for this purpose alone. There are clubs to buy guns through raffles. People stand with tin cups on street corners to collect coins for the purchase of hand grenades. Drives are conducted to acquire tanks. There are tag days for military aircraft. Cousin clubs sponsor dances to buy uniforms. The national governments simply neglect the problem of defense and let the burden fall on private agencies.

But the inadequacy of this system is apparent to all. People grumble that under such a policy there will never be a war.[2]

The questions we applied to Bono's comments may profitably be brought to bear on Polish's statement. To what extent do you find each vision inspiring… firmly based in reality, impossibly idealistic, too costly, or… ? Do you perhaps find yourself uncomfortable with both of these "alternate worlds" in the same way many resist being told that north doesn't have to be uppermost on any map, or that the "developed world" has no intrinsic right to its privilege? How do you see the two visions, and where do they fit into yours?

There's a Map for That!

Visions like these are – surprisingly – closely linked to maps. Any vision is in reality a projection in that it proposes, *projects*, an idea, an agenda for us to consider. And that, as we have stated, is precisely what a map does – it simultaneously operates on two levels: it (1) projects a curved surface onto a flat plane and (2) projects an agenda. The map sets before us a way of looking at the world. It is appropriate, therefore, that we now set world maps beside these and other visions.

If the world you want is lopsided: a system in which some dominate while others are short-changed, would not a Mercator look-alike represent that? If, on the other hand, your ideal world is founded on fairness (which might include one person, one vote; universal access to education, every country given its due), would not a Peters or Hobo-Dyer be consistent with that goal?

No map will magically bring about the world we want. But until we give our lofty goals visual expression, we'll work under a handicap. It's like running a race without ever seeing the finish line.

For that reason map projections like the Peters and its close relative, the Hobo-Dyer, pull us toward the world we want, serving like a magnet. And that, let us be clear, is not just a pleasant "extra" – it belongs to the essential nature of maps. What we're talking about here is nothing less than a radical revisioning: seeing things in a new way. The agenda, then, is to gain a new perspective on the world. As British geographer J. Corner says, "the function of mapping is less to mirror reality than to engender the reshaping of worlds in which people live."[3] We may not inhabit that reshaped world ourselves, but whoever said that's a reason to give up?

CHAPTER 9

The World We Get

"Between the idea
And the reality
Between the motion
And the act
Falls the shadow..."

T. S. Eliot

Mini-climates in Southern Ontario are friendly to fruit-growing; its vineyards look like a smaller Napa Valley; its peach orchards rival Georgia's. Growers rely on seasonal help. Thus, friends of ours regularly count on men from Jamaica to work their fields from spring to fall. Over time they developed a relationship of trust and friendship – more like family than a rigid hierarchy of bosses and workers. So, when the Canadian couple visited Jamaica they invited their worker-friends, plus their families whom they had never met, to enjoy time with them at their hotel.

Then the shocker hit. When the invited guests got to the hotel gate, guards brusquely turned them away. No apology. No explanation. Just one not-so-subtle message: this was a hotel for tourists; average Jamaicans were not welcome.

Let the story serve as a microcosm of our world. Privilege is not evenly distributed. Discrimination lives. Even with the best of intentions, things may turn out wrong. The world we may desperately want is not the world we get.

Between Us and the World We Want

Analytical critiques of our global malaise are never hard to find. Our purpose here is not to add to the existing mountain of weighty commentaries but to step back a bit, so gaining perspective.

One striking fact – at least to this observer – is how many analysts from widely varied backgrounds see a major split – a deadly serious fault line – in the world. They propose such terms as

- Third World
- Third and Fourth Worlds
- The Developed World, the Developing World, the Underdeveloped World
- Fast World/ Slow World
- The Core and the Gap
- The Core and the Periphery
- The Stuffed and the Starving
- Freedom's Divide
- North/South
- Viable States/Failed States

The terms themselves can be revealing. "Third World" gained currency during the Cold War. With East-West tension – centered on Moscow and Washington – demanding top attention and constant vigilance, suddenly inventing a term like the "Third World" introduced a whole new dimension. It called people to recognize that much of the world had urgent needs not related to the East-West power struggle.

Actually, the term first surfaced in French – *le tiers monde*. The common English translation loses part of the point, simply suggesting a race in which some people come in third. A more accurate translation would be the one-third world, which speaks to the size and importance of this segment. In fact some advocates, doing the math, conclude we should refer to *the two-thirds world*, based on sheer numbers of people.

When the North-South Commission, chaired by German Chancellor Willy Brandt, presented its groundbreaking report in 1980, it served to alert an international audience to that vast area it termed "The South." It found the symbol of its concern in the Peters World Map, and reproduced that map on the cover of its report, inking in a line running dramatically across the map to mark off North and South.

"East-West rivalry isn't the whole story," it seemed to be saying. "There is another way to look at the world, and we better not forget it."

Without denying the importance of the North-South split, *New York Times* columnist Thomas L. Friedman sees a world divided into tradition-bound societies and those eager to change. His book *The Lexus and the Olive Tree* offers an explanation of how some societies adapt to new realities: if the world wants cars or computers they will build them, even innovating and aiming for the highest quality they can achieve. Meanwhile, other societies cling to age-old habits, harvesting olives from the same groves that have stood there for centuries.

Should people who tend trees build luxury cars instead? *Could they?* What options do they have, really? Which points to another worldwide split: some enjoy freedom to choose; others live within severely limited options. This in turn hinges on educational opportunity, social expectations, access to resources, and living in an open, democratic society in contrast to a dictatorial regime. The gap is real, though we are learning that how to deal with it will never be simple.

Fig. 9-1 The findings of the North-South Commission gained wide distribution in book form through language or regional editions. The line separating "North" from "South" does not mean that everyone above it is privileged while on the other side all are oppressed. Certainly Australia and New Zealand, though south of the equator, are part of the "developed world." Some in the "South" are extremely wealthy; some in the industrialized countries live in Third World conditions. The marker serves to highlight prevailing conditions that make a huge difference.

Source: Akademische Verlagsanstalt

In a report to the United Nations, the Human Settlement Program predicted that by 2020 about 1.4 billion people would be living in slums. The world's present slum population is about 1 billion, most of them in the "South" or developing world. Without radical intervention by governments, 27 million more people will be forced into slums every year. That is the equivalent of all the people in New York's Manhattan plus all of Australia – picture these millions getting pushed off a precipice into poverty, no matter how hard they struggle; condemned to life without adequate shelter, without safe water, without working sewers, largely without hope.

With an estimated 90 percent of the world's slum dwellers geographically distanced from us, we may fall into the "out of sight, out of mind" syndrome. We feel sorry for them but seldom think about them. Nevertheless, warns *Forbes Magazine* in a 2007 analysis, "the world ignores slums at its own peril."

Economist Jeffrey Sachs is passionate about responding to the rich/poor divide. In *The End of Poverty: Economic Possibilities for Our Time* he points out that more than a billion people currently subsist on resources of less than a dollar a day. No matter how we try to explain it, that is *extreme* poverty. Sachs tackles head-on such common excuses as the idea that Africa is poor because it has corrupt governments. The root of the problem, in his analysis, is grinding poverty; deal with that and political improvement will follow. But first, he says, we may need to change our mindset – we in the West, the North, the developed world. We need to see Africa as the significant, vital part of the world that it is. That sounds like a clear rationale for replacing South-suppressing map images with the more realistic perspective of equal-area maps, letting Africa confront us anew in all its vastness. Wouldn't that seem a logical place to start?

Thomas P.M. Barnett, an adviser to the United States Department of Defense, selects "The Core" and "The Gap" as labels. Noteworthy is the fact that the former bears similarity to "the North" (or West or developed world) while "The Gap" is largely made up of former colonies now caught in the poverty cycle. And what solution does he recommend? Connectedness is key, he says: in a kind of extreme networking, nations caught in The Gap need to link up. Not only with other Gap nations but with the Core. Especially with the United States. Globalization – in communications, commerce, cultural ties, and military alliances – counts.

High on the list of effective organizations concerned about world poverty is the Millennium Project. Director John McArthur, speaking to the Canadian Institute for International Affairs, identified a 3-speed world. "The first speed is the part we live in, the rich world which has been blessed with an 'economic miracle.' The second speed has people living in extreme poverty, but with rapid rates of progress and social mobility. Then there's the third speed, which is more like no speed. These people are at the bottom of the ladder. Actually, they're not even on the ladder. These are the people living in parts of Southeast Asia, Northeast India, Africa and Central Asia, who die 'silent, voiceless deaths.'"

Some analysts pay more attention to cultural issues than to questions of affluence and poverty. In his provocative book *The Clash of Civilizations and the Remaking of World Order,* Harvard political scientist Samuel Huntingdon predicted a war of values. While the clash may be interpreted narrowly as between Islamic jihadists and Christian fundamentalists, it may be seen as having far broader implications. "Not just youngsters, but people with families, in their 30s, are willing to go to Iraq and blow themselves up. That is something new," points out Reuven Paz, a specialist in contemporary Islam who is based in Israel. Paz sets such nihilism alongside the American policy of "endless war against terrorism," exemplified by the invasion of Iraq and other countries. Thus these two thoughtful observers from very different vantage points speak of a single reality: a global fault line that kills effective communication as it sets up "righteous" battles against "evil."

While many seem surprised by the vehemence and vigor of the resentment on the other side of the divide – whether perceived as the poor, the South or the Islamic world – some perceptive analysts saw that over 50 years ago. At the height of the Cold War, Reinhold Niebuhr observed that

> The victors [in the Cold War] would... face the "imperial" problem of using power in global terms but from one particular center of authority, so preponderant and unchallenged that its world rule would almost certainly violate basic standards of justice... If [we in America] assume that all our actions are dictated by considerations of disinterested justice...the natural resentment against our power...will be compounded with resentment against our pretensions to a superior virtue."[1]

Niebuhr, who taught at New York's Union Seminary, was speaking about American power, but surely it requires no stretch of the imagination to see the resentment Niebuhr observed in 1960 evolving into "revenge" and spilling over to countries aligned with the USA – say the G8 counties, the North, and the developed world.

Resentment of American power becomes the focus for other astute commentators as well. William Appleman Williams spoke of U. S. architects of the postwar world as operating out of "visions of omnipotence." As in physics, where an action calls forth an equal and opposite reaction, so in human affairs: such over-reaching pride creates a climate of fighting back. Williams was one of the 20th century's most influential analysts of American foreign policy.

Edward Peck, former U. S. Ambassador to Iraq and deputy director of President Reagan's task force on terrorism, asks, "Why is it that all of these people hate us? It's not because of our freedom. It's not because Britney Spears has a belly button or because we export hamburgers. They hate us because of things they see us doing to their part of the world that they definitely do not like."

From a Muslim writer living in the West we hear, "Muslims are in a state of crisis, but their most daunting problems are not religious. They are geopolitical, economic and social – problems that have caused widespread Muslim despair and, in some cases, militancy, both of which are expressed in the

religious terminology that Muslim masses relate to." Haroon Siddiqui, a past president of PEN Canada[2], points out that most Muslims live in the developing world with its still-recent history of colonialism. That is not the only source of their difficulties, but it is one significant factor. He provides statistics:

> The total GDP of the 56 members of the Islamic Conference, representing more than a quarter of the world's population, is less than 5 per cent of the world's economy. Their trade represents 7 per cent of global trade, even though more than two-thirds of the world's oil and gas lie under Muslim lands.
>
> The standard of living in Muslim nations is abysmal even in the oil-rich regions, because of unconscionable gaps between the rulers and the ruled. A quarter of Pakistan's budget goes to the military. Most of the $2 billion a year of American aid given to Egypt as a reward for peace with Israel goes to the Egyptian military.
>
> The most undemocratic Muslim states, which also happen to be the closest allies of the U.S., are the most economically backward.[3]

> "Convictions matter. At least our own convictions – the affirmations, commitments, and practices that are central to our personal and social identities – matter to us. Yet because we live in a world of unprecedented global interaction, the convictions of people everywhere matter to all of us whether we know it or not."
>
> George Rupp, CEO of the International Rescue Committee

If the West sees itself as under siege, whether by Jihadist Muslims in particular or by the world's downtrodden in general, it may be time to ask whether "they" may not also feel threatened by the West. In the case of the Jamaicans not allowed past the guards into a hotel in their own country, they seem to have taken a forgiving attitude. But can we always count on tolerance when people feel they bear the brunt of injustice? If today some Muslims lash out at what they perceive to be their "oppressors," what guarantee do we have that other groups will not join them tomorrow?

If the world were a family – persons in one room scrounging for scraps; in the next room a banquet being served; in one part of the house people dancing and enjoying a party while just out of sight other family members lie sick and in wracking pain – would we not label the family dysfunctional? Would somebody not call for professional help? In the kind of world we actually get – so different from the world we want – the follow-up questions become, "Where will healing come from?" and "How can we begin to make things more fair?"

Meanwhile, the fault line running through our world, no matter what label we choose to give it, and the resulting siege mentality on both sides, has consequences. And these threaten, like some rushing tsunami, the well-being of the world.[4]

The gap has another dimension as well. It demonstrates the disconnect between our pious declarations and our actions, between our self-image of caring about justice and the reality of an unbalanced world.

While the chasm of separation would present a problem in any period, it takes on special urgency in our day. A century ago the most impoverished knew little about life in the developed world; today what happens "here" can be instantly known "there." Consider the operators in a call center in Mumbai, India, devoting their working day to business transactions on behalf of clients in the North – do we really suppose they are unaware of the impassible divide between the fabulous funds they move electronically, and the aching hunger of children they see when they go back home? Thomas Friedman declares in *The World Is Flat*, "They do not want our Nikes or our Blackberrys; they want our jobs, our kind of opportunity." Strong feelings within the impoverished world are not to be dismissed. No matter how we label them – jealousy, ambition, frustration, envy, hope, longing, impatience, the sad relic of long-ago injustice – the reality is they form a volatile mix.

This is not to claim that the North-South split is the only fact we need to address to understand the

world-as-it-is. But it is to assert that only by being fully aware of that reality can we be realistic about the world we've got. So let's focus on a seldom-asked question about how borders function in the global community.

How Long Can Borders Last?

Somewhere in the world a medical team is being dispatched to a land most of us will never visit. Borders that cut people off from others virtually vanish before the medics' single-minded determination to help victims of disease or disaster. "Borders? They're challenges, sure, but in the final analysis they're just lines on a map; they have no reality except what people give them." Any wonder these people are called Médecins sans frontières/Doctors without Borders?

Other examples similarly instruct. The European Union represents a high-minded effort to break down walls of insularity built up over centuries.

Some things do not change, however. Doctors serving beyond borders are still doctors; the mapline between France and Germany has been downgraded in importance, but people on one side still speak and think in French; on the other, in German.

Would a world without borders still be the world? Would nations without borders still be the nations we know? Is a world without borders even thinkable?

Realistically, no. Border-free is not the world we get. Still, there are some factors – gentle pressures, if you will – trying to move us in that direction, and some signs that they are having some effect. Here are some examples, with no attempt to rank them:

- The European experiment. True, the EU has its problems, but who today contends that the continent should return to its former status of xenophobic fiefdoms? Would anyone choose either the map or the mindset of the continent that gave us modern nationalism, imperialism, and the Holocaust? The clear consensus is that letting national borders fade to a less prominent role – on the map and in trade and commerce and freedom of movement – represents real progress.

- The image of Earth from space. It's obvious from this NASA photo that every border on the planet has been contrived by human beings. And it's clear from history that many a border marks ancient scars and continues to be imposed and defended by force. Therefore, logically, borders can be scrapped should we ever muster the will.

Fig. 9-2 Earth from space
Source: NASA

- The Appalachian Trail, which ran for years from Georgia to Maine and now reaches into the Maritime Provinces and Newfoundland, may be further extended – get this! – to Portugal, Spain, even Morocco. Those European-North African countries surprisingly have mountain ranges that owe their origin to the same collision of tectonic plates that formed the American Appalachians. Some 300 million years ago they were all part of the same mountain mass. Then, about 100 million years later, the large landform known as Pangaea split apart, so some of its mountains are now located in Europe, some in North Africa, some in North America. Appalachian activist Dick Anderson explains the motivating vision: "We want to get people to think beyond borders." He sees this as a way of helping people connect across cultures.

- Electronic-communication systems now bypass borders as if they weren't even there. Who cares whether your email correspondent or Facebook friend lives under another flag? Even in those cases where fearful governments try to limit access by erecting high barriers, the trend would seem to be toward people exercising freedom to speak and act on their convictions. The full change may not come tomorrow, but no borders can match the power of the human spirit coupled with the electronic revolution. If members of Facebook were to consider themselves a country their population would be the third largest on Earth (800 million).

- Poet Robert Frost gave the world a great insight when he spoke of a "something" that doesn't go for walls – that in fact stubbornly resists all bordering off. He never defined that "something" – perhaps no one can. Like the breeze we never see but know only by its effects on waves, leaves, and skin, we experience that pervasive force when it helps bring down the Berlin Wall or the fortress walls of Quebec or ancient Troy.

- What are borders for, anyway? They are there to *define*. They make clear what belongs inside the line. They do this very largely by keeping "the outside" out, whether that is defined as foreign people, products, or ideas. In times of peace nations may defend their interests by tariffs on imports; when things heat up they escalate troop levels. To what extent are the borders marked on maps the result of a deep-seated parochialism, and to what extent do they encourage it? Would the territorial imperative – "Keep out; you're not welcome here!" – be any less operative if there were no borders? What, indeed, do we mean when we say we "belong" to a certain country – in what sense does an abstract entity *own* us? Is it possible, given our history, our group loyalties, our maps, to conceive a world where loyalties are broader than the nation?

Still, reality is never simple. If we quote Frost should we not also remember his observation, "good fences make good neighbors"?

Shift the debate: are there some borders that should be abolished while others are beefed up? Look at any political map, that is, one that shows nations and borders. Which lines would you label beneficial and which would the world be better off without? How about the U.S.-Canada border? America's border with Mexico? The line between England and Scotland? The barrier between East and West Germany, including the Berlin Wall, that fell in 1989? The DMZ separating North Korea from South? The line between the developed world and those nations caught in the cycle of poverty?

How to Respond?

If you're reading this book you have three major advantages. First, you have the discretionary time and money to buy and read this book (or you have access to a local library or your professional association gave you a copy as a member benefit). Second, you live in a free society, which sets options before you. Third, you're aware of the problem. Millions aren't. Or don't care. That means people like you and me can take informed action. Consider how the following "starter" ideas might make a difference:

- By linking with an existing group or starting a new one, set up a study-action program around issues this book has raised.
- Support an organization that focuses on world affairs or justice, perhaps in your community, synagogue, church or mosque, library or school.
- Find out about book clubs in your area; suggest this issue as a focus; offer to participate.
- Remember that "Third World" need not be only overseas. Events such as Hurricane Katrina and the recession of 2008 (and following) reveal that pockets of poverty are hidden within even the wealthiest nations. Some may be near where you live; seek them out and explore how you can support those neighbors.
- Do an inventory of activities in your community that engage in intercultural dialogue. Get to know people with a different background or outlook.
- Sponsor an intercultural experience. A Toronto church does this at Thanksgiving, enabling "new Canadians" to experience this uniquely North American holiday, tell their stories, and contribute to intergroup understanding in the city that the UN has labelled the most cosmopolitan in the world. In this case hundreds of immigrants from Central America, China, Nigeria, and other countries have

"I can see being proud that your kid watches the news. I can see being a little proud that they understand they have privileges in this country that other people do not. I can see being a little relieved that they know not everyone goes to bed with a full stomach, that they can at least imagine the fact that war causes unimaginable pain. But then what? The punch line from the religion of gratitude: "We're so lucky that we live here instead of there." Really? That's it? Never been prouder?

What's missing from that worldview – and this is no fault of the teenager – but what is missing from that worldview is the perspective that you might get in a Christian community that would take you from lucky to actually doing something about it."

This comment* is from a specifically Christian perspective. How do you see the author's perspective applying to faith communities of other traditions? Do you agree with the writer that feeling "lucky" or even thankful is not enough? Why or why not? How can participating in a faith-based or other community help you make a difference in the world? Alternatively, can we look at the question through the eyes of a Third World community striving for get out of poverty? What difference does it make to you where your development support comes from?

*Excerpted from "You Can't Make This Up!" by the Rev. Lillian Daniel, in *Christian Century*, Sept. 1, 2011, slightly adapted.

formed friendships with long-term residents. Pushpins on a Peters map, identified with people's names, mark the locations from which they come.

- Investigate the possibilities of "twinning" your community (city, town, church, school, for example) with a comparable group on the other side of the global divide. Help may be found at www.sistercitiesinternational.com among other sites.
- Find out more about NGOs (non-governmental organizations) and their role in responding to human need. Reflect on the fact that some observers see these groups' humanitarian work as one of the most hopeful developments of modern life. For example, Brian Stewart, a veteran foreign correspondent, says that the modern NGO movement is the fastest growing sector of the world economy. "Everywhere I've gone in the last ten years, I've been struck by the way [they] form a 'critical mass,' bringing better education, health and respect for human rights to everspreading circles of people."
- Check with your representatives in Congress or Parliament; what stance do they take on issues of fairness locally and globally? Listen, encourage, push, write, or lobby as needed.
- Seek out persons and organizations working for fairness to all[5] and peace (examples abound, including fair trade enterprises such as *Ten Thousand Villages* stores, or *New Internationalist*,

co-publisher of the original e-book that preceded this paperback edition); ask how you might help or volunteer to support them.
- Deal with the challenge of narrow worldviews. If the first requirement is to change people's mindset, how do we go about that? If you see equal-area maps such as the Peters and Hobo-Dyer helping that happen, how can you help make them more widely known?

Are the world's problems subject to human control? Can hope survive? Getting our daily fix of bad news may push us to say "NO." But that may be precisely the time to celebrate the "YES!" – those moments when people refuse to surrender to despair. Celebrate Gerhard Kremer: can you visualize him watching, feeling the pain of families saying tearful goodbyes to men setting out to sea, knowing they might never see them again? So he created a map that changed all that. Celebrate Arno Peters: he saw a lopsided world; instead of keeping silent he spoke out for justice through an equal-treatment map. Celebrate medical mappers and weather mappers and eco-mappers who make life better for us every day. Celebrate all who take action for a fairer, more liveable world. Celebrate hope with song and vivid art and joyous dancing!

The point is surely clear: the world we live in is not the world we want. But there is good news: we can all do something to narrow the gap. That, some of us are convinced, directs us towards a world where everybody wins.

I bike. On a favorite trail I regularly clock one segment at 15 minutes. Later – funny thing – I do the opposite direction in just 8 minutes. The secret? Nothing to do with being more experienced; I'm simply going downhill. But nobody can roll down until they've climbed up; we all learn that only through the pain can we claim the gain.

I offer no apology for laying out the problem in perhaps depressing detail: developing awareness is

always the first step. Now it's time to shift focus. First we'll look at the surprising, little-understood-yet-significant connection between maps and faith systems or, more broadly, the values that claim us. Then we'll move on to investigate how our values and our maps together can make a world of difference. To those changed perspectives we now turn.

CHAPTER 10

Maps and the Faith/Values Connection

"Values will be the most important political question of the 21st century."

Jim Wallis, Christian social activist

Lance Armstrong, as we all know, has problems. But long before he confessed his wrongs on TV, he got one thing right. "It's not about the bike," he said.

"Yes, Lance," we might add, "it's not about the bike. And it's not about wearing the yellow jersey or hearing the crowds roar their approval. It's about self-discipline and skill; it's about fairness. Integrity. In the final analysis, it's all about values."

Similarly, Arno Peters got it right; analyzing the debate over his map he declared, "It's not about a map – it's about a worldview."

> Economic strength and military force will never build peace in a globalized world, said Lech Walesa, former president of Poland, at the University of Central Florida. Human values are paramount. And the developed world, especially the United States, must be a power in renewing those values.

To talk about worldviews is to move beyond, though never to reject, the domain of mathematics – the realm of projection systems and rectangular vs. curved grids – and enter the realm of attitudes, relationships, how we connect with one another and with the planet. It's to shift from the hardware to the software of our inquiry. In a very real way it is to enter the realm of faith and values.

Note the language: we do not claim it is to enter the realm of "religion" as that is often understood. In fact, you don't have to be "religious" to like the Peters map. This is exemplified by Peters himself; he was highly skeptical of all formal religions.

Yet without the support of religious organizations this map would never have achieved the worldwide recognition it enjoys. Faith communities were early enthusiasts and remain strong supporters. Is this just coincidence, or is there something more going on here?

A helpful way to get at the question is to focus on values. When we move beyond paste-on labels, getting through to those motivators that drive our daily behavior, it's no longer a question of whether you are Baptist or Catholic, Muslim or mystic, Hindu or humanist. If you're Jewish it's not a question of whether you keep kosher and strictly observe the Sabbath. What does count is what your goals are, what springs you into action – in short, your values. That holds true whether that "you" refers to you as an individual, or to a family, an organization, or a nation. While we will focus in this chapter on the Peters map as the most dramatic example of the principle, the same values analysis may be applied to other maps. What worldviews are implicit, say, in the map that sees Iraq primarily as a source of oil for the United States, the map that shows the world centered on Havana, then enlarges Cuba (see Chapter 11), the map that flaunts the British Empire in grandiose glory?

An early proponent of the Peters map in North America was the Evangelical Lutheran Church in America. So strongly were they convinced of its ability to reshape people's operative image of the world that within months after the map was published they distributed many thousands of copies. The Rev. Arthur Bauer, then with the church's Board for World Mission, explained it this way:

> The Christian message includes an emphasis on justice for all, based on the love of God which is extended to all. That message cannot utilize a map that sets forth an inaccurate, distorted world view. The Peters map appears to be the best educational tool for showing us our place on Earth. The values and purposes of this map coincide well with the teachings of the Bible and the church.[1]

The Roman Catholic Church emphasizes what it calls "the preferential option for the poor."[2] To know what this means, consider three statements – more precisely a fact, a principle, and a recommendation:

- First (and most obvious): some people in the world are poor, marginalized, and oppressed.
- Second: people of faith have the duty/privilege to respect those marginalized people.

- Followed by the injunction: merely to deal with them on an equal basis is not good enough, for that only perpetuates their disadvantage; "second-mile generosity" is called for in order that accumulated wrongs may be righted and fairness achieved.

It may come as no surprise, then, to learn that Pope John Paul II had a large Peters map hung in the halls of the Vatican. To all who saw it, it became a wake-up call to the importance of the Two-thirds World, and the need to keep marginalized people and nations constantly in the forefront of attention. The United States Catholic Mission Association likewise gave the Peters map pride of place, starting in the late 1980s, in their public materials.

In the United Kingdom an early version of the Peters map was distributed by Christian Aid, complete with a quotation from Psalm 89, from which they took the map title, "North and South."

The World Council of Churches, encompassing major Protestant, Anglican, and Orthodox bod-

Source: Akademische Verlagsanstalt

CARTER CENTER ACTIVITIES BY COUNTRY

The Carter Center has helped to improve the quality of life in more than 65 countries.

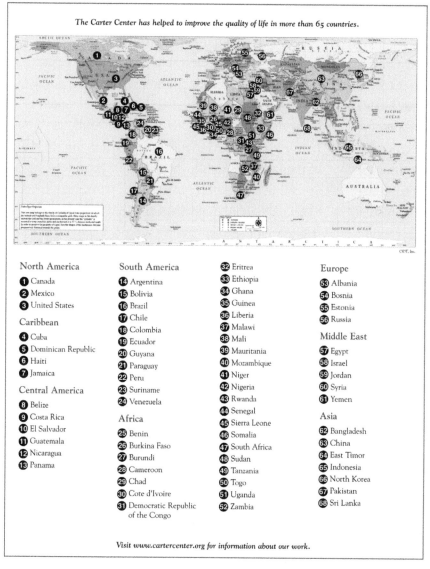

This world map, known as the Hobo-Dyer, received favorable attention in Oslo, Norway when the 2002 Nobel Peace Prize was awarded, and later when it was displayed at the Carter Center in Atlanta, Georgia. The Carter Center effectively serves needy people in 78 nations.

North America
1 Canada
2 Mexico
3 United States

Caribbean
4 Cuba
5 Dominican Republic
6 Haiti
7 Jamaica

Central America
8 Belize
9 Costa Rica
10 El Salvador
11 Guatemala
12 Nicaragua
13 Panama

South America
14 Argentina
15 Bolivia
16 Brazil
17 Chile
18 Colombia
19 Ecuador
20 Guyana
21 Paraguay
22 Peru
23 Suriname
24 Venezuela

Africa
25 Benin
26 Burkina Faso
27 Burundi
28 Cameroon
29 Chad
30 Cote d'Ivoire
31 Democratic Republic of the Congo

32 Eritrea
33 Ethiopia
34 Ghana
35 Guinea
36 Liberia
37 Malawi
38 Mali
39 Mauritania
40 Mozambique
41 Niger
42 Nigeria
43 Rwanda
44 Senegal
45 Sierra Leone
46 Somalia
47 South Africa
48 Sudan
49 Tanzania
50 Togo
51 Uganda
52 Zambia

Europe
53 Albania
54 Bosnia
55 Estonia
56 Russia

Middle East
57 Egypt
58 Israel
59 Jordan
60 Syria
61 Yemen

Asia
62 Bangladesh
63 China
64 East Timor
65 Indonesia
66 North Korea
67 Pakistan
68 Sri Lanka

Visit www.cartercenter.org for information about our work.

ies and through them representing some 560 million Christians around the world, regularly uses the Peters in its publications and educational resources.

President Jimmy Carter of the United States, when he received the Nobel Peace Prize, deliberately chose the Hobo-Dyer (as we pointed out earlier, it shares with the Peters its foremost quality: a totally accurate and fair treatment of all areas, all nations) to present the worldwide work of the Carter Center. While there is no overt faith reference here, Carter's personal convictions are well known. His own value-system and that of the Carter Center find clearest expression in a world map that treats all nations with respect.

In the days before the Peters map had entered either the commerce or the psyche of the English-speaking world, it was a German publisher who brought it to my attention. *Evangelische Weltmission,* the enterprise he represented, was distributing it primarily to churches to support their work in building global awareness and action. The map spoke to the need to treat all peoples fairly and with dignity.

In France, *Cimade,* an ecumenical activist organization focused on justice for marginalized people everywhere, makes the Peters its map of preference, offering a realistic view of the world as it is and the world we work toward. Its collaborator, *le Comité Catholique contre la faim et pour le développement* (Catholic Anti-Hunger Development Committee), developed its own tools for elementary-level education about the Peters map and how it stimulates awareness of what we call developing nations or the Third World.

Speaking for the World Association for Christian Communication, General Secretary Randy Naylor says,

> The Peters is prominently displayed in our London offices; it will be even more arresting when we supersize it for our new Toronto location. The Peters, as we use it in a marked-up version, makes it clear that communications in practice – global news, documentaries, drama and entertainment – often reflect a privileged and extremely narrow political, economic, sociological and religious worldview. The stories of the world represented by the South on the Peters map are integral to a fair, just and holistic worldview. Without such stories and perspective we not only shortchange others, we shortchange ourselves. The Peters helps us identify the hope of living in this world that combines great need with great potential.[3]

The United Methodist Church, prominent among Protestant denominations in the United States, placed a Peters map of the world etched in glass at the entrance to its Global Ministries office. You can't miss it – not just because it stands more than six feet tall, but as you walk into the offices – literally *through* that map – you get the sense that you're part of a "different" world; you become a participant in a mindset that takes the needs of the whole world seriously.

Canada's largest Protestant body, the United Church of Canada, so identified with the goals made visible in the Peters map that it placed a large order customized with its own imprint. Mardi Tindal, who had helped launch the map by interviewing me on TV, became the 40th Moderator of the denomination in 2009. She comments:

> As a learning tool the Peters Projection map has helped people, including those in the United Church, to think about the assumptions that shape how we perceive the world. That "ah hah" moment when we begin to see the world from a different perspective is a crucial step in becoming involved

in justice making. The map and its use in educational workshops has been part of the United Church's commitment to decolonize its worldview and therefore the way we participate in mission. These types of learning tools continue to be crucial to the church's ongoing commitment to live faithfully in the midst of empire and to becoming an intercultural church.[4]

The Mennonite-sponsored fair trade retail stores, Ten Thousand Villages, have sparked community outreach programs on global justice. Using both the Peters and the What's Up? South! (see Chapter 2) has been effective for them; in fact, these maps have often been among their ten best-selling items sourced from outside their traditional supply chain.

Additional comments from a faith perspective may be found on the DVD, Arno Peters: Radical Map, Remarkable Man.

Malcolm Fife, a Canadian, writes from the vantage point of Lifeline International, an interfaith network of volunteers working to assist persons in crisis:

> When I first encountered the Peters map and studied the related Handbook I was propelled into a new way of relating to people everywhere. I soon found a practical use for this understanding. As active members of Lifeline, my wife, Fran, and I worked up a customized map of the world to highlight those places where Lifeline operates – from Papua New Guinea to South Korea, Australia, Botswana, the USA and beyond – and showing where the more than 600 delegates to our Toronto world convention came from. The Peters was perfect for our purpose. The resulting map was so large it clearly dominated our meeting space, became more than decoration, more than an educational device; it forcibly reminded us, as one overseas participant remarked, that in spite of cultural differences and the controversies that threat-

ened to tear us apart, we were all one, in one world with one purpose.[5]

While many a map might be utilized to highlight the unity Fife and his colleagues are committed to, the Peters did the job especially well because it helps viewers visualize equal respect for all nations.

Stephen Goldstein, an American with two decades of experience preparing missionaries to serve in other cultures, points out that "missionary policy understands mission to be not from North America and Europe to the Southern hemisphere, Asia and the Pacific; rather, we celebrate the fact that our missionaries come from everywhere and serve everywhere. It can be no surprise that in orientation and training we have relied on the Peters map to support this worldview."

In the parish or congregational context, examples abound of people finding the Peters map helpful. In an unsolicited comment a Baptist minister writes,

I use the Peters map as part of my Sunday School and mission teaching... [T]he Peters... has been extremely helpful in making the point about the misconceptions people (especially Americans) have about the world....

By the way, after dealing with the visual contrasts between Mercator and Peters, I usually get into an issue that flows from that: HIV/AIDS, bringing out the point that while it is a relatively small issue in North America, it is a huge issue in sub-Saharan Africa, yet perceptually people in the USA have little sense of urgency about what's happening in Africa.

This becomes a launching pad for our mission agenda: getting the congregation committed to the needs of the two-thirds world.[6]

Whenever I do map presentations with Jewish groups, I am impressed with their openness to questions of global justice, which is never far from a consideration of equal-area maps. So regular an occurrence cannot be simple coincidence; there has to be a connection. Here is what Ellen Hofheimer Bettmann, an anti-

bias educator with the Anti-Defamation League, has to say:

Jewish teaching includes the concept of *tikun olam*, a mandate that means repairing the world; this theme underlies many rabbinic teachings. While the concept may sound vague, we Jews are helped in the process of figuring out what to do by the concept of *mitzvot*, loosely translated as good deeds. There are 613 *mitzvot* in Judaism, biblical in origin. One of these, central to the lives of social justice activists, is "Do not stand idly by" (Leviticus 19:6). Noticing suffering, caring, and taking action is central to Jewish life. Another key element in Judaism is remembering our history as slaves and the exodus from Egypt. This remembering, which is another of the 613 *mitzvot* and forms part of the *shema* or daily prayer, compels us to work always to ensure that all people are treated fairly and justly. This distinctly Jewish perspective is a major reason why I am such an enthusiast for maps like the Peters and the Hobo-Dyer.[7]

Has Buddhist philosophy been threaded into Arno Peters' design? Not intentionally, but the compatibility is remarkable. A Buddhist expression says, quite simply, "What we think, we become." And Thich Nhat Hanh wrote, "...change in the world begins in personal, inward change." Those who experience a new view of the world, stimulated by the Peters map and opening themselves to its full implications, enter into an inward change that leads to change in the world. "Enlightenment" may be private, but it can bring public results.

The Chinese sage Confucius, whose perspectives on life remain influential 26 centuries after his death, held that always treating others with respect was of highest importance. In a book on maps it is appropriate to ask, "Which maps show greater respect: those that cut some countries to a fraction of their size, or equal-area treatments?"

Source: Global Citizens Network

High values shared by the world's faiths and by many who identify with no religion come through clearly in the mission statement of the Global Citizens Network. "The people of the world are one people, enriched by differences, united by the common bond of humanity," they affirm. That perspective plays out as international teams of volunteers build bridges of understanding and respect between peoples. The organization commissioned the world's largest Peters map for its Minneapolis headquarters – 16 feet wide! Comments staff member Laura Kurland,

> It is the best icebreaker we could have ever asked for! My favorite story about this so far was when a delegation of Latin American civic society leaders arrived in our office. When we took a break during the meeting they all ran over to the map and discussed their observations about the proportions and sizes. It was an absolute hit!

Inclusive/Exclusive Points of View

But let's be clear that not all expressions of religion, not all value systems, endorse the Peters or any similar map. There is, in fact, a deep divide, with some on one side, some on the other.

First, the context. Ever since history began people have been persecuted or even mass-murdered because of the values they held. Just wearing – or not wearing – a certain religious label has, in many cases,

been reason enough to get a person tortured or killed. Early Christians in the Catacombs, Jews in Nazi Europe, Muslims and Eastern Orthodox Christians during the Crusades come quickly to mind. Sometimes the driving force for hatred focuses on ethnic or political labels: Joseph Stalin came close to wiping out the Kulak peasants because they owned property; 16th-century Spanish conquistadors effectively destroyed the Inca civilization of Central America; suicide bombers – Islamic extremists – murdered some citizens of the Western world who happened to be in the wrong place in New York or Shanksville, Pennsylvania, or Washington, D.C., on September 11, 2001; or in London on July 7, 2005 or Mumbai on November 26, 2008.

To that horrible list we must, unfortunately, add a current example. But it's of a different sort:

- This one does not record a killing that *has* taken place; rather, it glorifies the idea of a mass killing that, it predicts, *will* take place.
- This kill-off dwarfs all others we could list. If the others were mass, this one is "masser."
- The agent in this case is not people out of control but a vengeful god. Or, as some would have it, God.

A case in point is the "Left Behind" series of thrillers. While fictional stories, they do reflect a value system and therefore need to be taken seriously. The "Left Behind" books have sold more than 65 million copies worldwide, and have long been bestsellers in the United States, and been adapted for movies. In the story *Glorious Appearing*, Jesus returns to earth to wipe out all non-Christians. The authors describe a vast sinkhole opening up in the earth, swallowing all who do not hold the "right faith." Their cries of despair reach deafening decibels. Even so, those who look on stretch out no hand in help. Then the Earth closes up again, and all is silent. The action? Violence. The purpose? Exclusion. The basis? Totalitarian intolerance.

Are you shocked by the ethnic cleansings of the 20th and 21st centuries? Mere practice runs compared with this act of total terrorism! Yet underlying all the examples just mentioned is one reality: how surprisingly similar they are. They all reject fairness and acceptance; they all opt for self-righteousness and "us first."

Neither Hitler nor Stalin nor the colonial masters of the world ever experienced the attitude expanding perspectives of today's many maps. Their schools never featured an equal-area treatment, nor even a thematic map showing levels of poverty, illiteracy, or military expenditures. That, of course, is history; what is more relevant today is that those who hold the most intolerant views in our time are no fans of the Peters or the Hobo-Dyer or any other map that promotes fairness for all people.[8]

> Neither lopsided maps nor intolerant religious perspectives are about to be nominated for a Nobel Peace Prize. Wrong views are no less wrong when they are perpetrated in the name of cartography or of the gods we worship. Today's interdependent world demands better.

Let's face it: it's easier to consign – in a novel, in one's mind, or in practice – whole groups of people to death or unending torture when one regularly sees uncounted millions – no, make that billions – in a position of inferiority on a map. Conversely, wherever an equal-area map affirms the essential equality of all peoples, respect and empathy gain visual support.

How Wide the Circle?

In Chapter 2 we made the point that we all exist at the center of our worldview. It can't be otherwise, simply because we can never jump out of our own skin or see through any eyes but our own.

The question here is not whether that's good or bad but how wide our world stretches beyond our kind of people, our interests, our square meter of Earth. The function of faith and the function of good global representation – as we understand the terms in this context – effectively merge, because both exist to stretch our worldview, to enlarge the circles of our concern. Students of human behavior often analyze relation-

ships as being "in-group" or "out-group." Visualize a circle with you at the center: those inside are the OK types: We accept them, feel good about them, make common cause with them. They are "us." Everybody outside the circle, on the other hand, is "them" – different, less important, sometimes demonized. In their presence we may feel vaguely uncomfortable, even threatened.

On what basis do we draw the line between "us" and "them"? In the example just cited from *Glorious Appearing*, the difference between the "saved" and the "damned" – the "goodies" and the "baddies" – is seen in religious terms, but other factors can function as effectively. People draw lines based on skin color, economic class, sports enjoyed or teams supported, management/labor, political preference, geography, profession, accent, what gets covered by clothes and what doesn't, educational level, lover of classical music vs. rock or hip-hop... the list goes on and on.

> "Maps help to tell us not only where we are but also how we view the world. How we view the world, in turn, defines who we are and from where we came."
>
> David Woodward[9]

Do you see the connection? Minimizing the worth of others and minimizing their place on a map are not the same – but they are strongly correlated. *Visualizing* others as less, and *treating* them as less must be seen as two sides of the same ugly coin. Seeing a map never engages just the eye; it involves the brain and, beyond that, how we process, intellectually and emotionally, the information we receive. Seeing the world as divided between "us" and "them," with *us* being superior and *them* inferior, can take many guises. When given "religious" expression it begins to look like the "saved for heaven" and "damned to hell" split. The same mindset expressed on maps can look a lot like a Mercator-based, North-biased map of the world.

The Values Gap

What we're dealing with here is a sharp discontinuity of values. "Some people are important": that's a value that seems to be universally held. Are others also important, or are they innately inferior? If important, how important? In the final analysis, do those in the out-group count as much as those in our in-group? Do all people merit respect, or do some make us hit the *REJECT* button? Are relationships to be built on mutual respect? Or by we dominate/you submit; we exercise power/you follow orders?

Those who find the politics of rejection and put-down intolerable, whose values support a fairer world, tend to like what equal-area maps stand for. Arno Peters didn't invent fairness or concern for others. Neither was he the first to create an equal area map. But he linked those concepts and gave them visual form. Through his map the sometimes abstract idea of "justice" becomes visibly real. The most natural thing in the world is that those who stand for justice should opt for the perspective that Arno Peters brought dramatically to popular attention.

Maps as Semi-Sacred

Whatever other functions faith-systems serve, they strive to express meaning in human existence. Whether the focus is on the Hebrew prophets or the parables of Jesus, on ecstatic union with the divine through dance or being caught up in awe while watching the sun set over the distant horizon – or a thousand other possibilities – the many faiths of humankind have in common the human hunger to find meaning in life.

Maps are part of that profound search. That is, maps are more than squiggles and dots; maps are stories, and in those stories lies meaning, and in that meaning we can discern direction and purpose.

This is not to claim that maps rise to the level of "the holy;" they will never call forth the response associated, for many, with the Star of David or the Cross or the laughing Buddha. Nor, for that matter, with the Stars and Stripes. But in a pragmatic way any faith-system is free to utilize secondary symbols as well as those it considers most sacred. Since maps are always overflowing with meaning, I assert that some may be said to overlap the realm of religion. They have an innate connection with faith-systems.

Perhaps that connection, that territorial overlap, that ability to look at the world-as-it-is and perceive a world-as-it-can-be lies at the root of the faith/values response to equal-area maps. Arno Peters correctly understood that rejection of his map was not primarily a technical, cartographic critique but a problem with his worldview. Conversely, the positive response to his map by faith groups and values-focused movements reflects their recognition that in his map they have a strong ally. The seemingly disparate worlds of expressing faith and making maps come together in a mutually reinforcing way. The challenge now is to see how maps and values affect each other, and how both have power to change the world. Maps and faith? To some it still seems an improbable marriage, but I assert that they are made for each other.

That is the faith/values connection.

"The unexamined life is not worth living"

Socrates

By the same reasoning we may say that anything – yes, anything – that pushes us to examine the meaning of life is worthwhile. Accordingly, it may be time to appreciate maps for their uncanny power to shake us up, to disturb the status quo, to question what is.

CHAPTER 11

Why Care?

"We couldn't live without maps. All our dealings with the world stem from how we create them."

Hooley McLaughlin, Director, Ontario Science Centre, in Canadian Geographic Annual, 2000

Our question "Why Care?" is really two questions: "Why care about the world?" and "Why care about maps?" But here's the point: the two are profoundly inseparable. To exercise an informed caring for the world is to recognize the importance of geography and maps. To understand maps in all their richness is to move beyond tangible documents to the world they point to, the living arena of our planet and its people.

Why Care about the World?

Sometimes cataclysmic events exercise such fatal fascination for people that other concerns retreat to the margins. When health problems hit – or financial crises or terrorist attacks or natural disasters – it becomes very difficult for even the most public-spirited citizens to focus on distant needs. Who can concentrate on how to get to the world we want when the world we get saps all our energies?

But if the world as it is makes it harder to see clearly the world we want, it never fully wipes out our hopes. The goal of a better world lives on in spite of all setbacks – otherwise the vision would long ago have vanished. Its tenacity underscores why the appeal to fairness, for justice and peace and service and high ideals make some sort of stubborn sense in spite of the contrary forces of cynicism, selfcenteredness, and "who-gives-a-damn."

Catastrophes can, amazingly, lend new urgency to the need to care deeply for the world. If the world is to be a better place – to make any progress toward the ideal we outlined a couple of chapters ago – freedom from fear must displace our strength-sapping concern for terrorist threats, food on every table must win out over hunger and, above all, at least roughly equal opportunity must replace the widening chasm between privileged and exploited. Times of crisis can even provide a "seize the moment" opportunity, since whenever old systems break down there is room to develop new ways of doing things.

We're not there yet. But some say they see a way toward the goal. Get a different perspective on the world – that is exactly what we need, according to geographer-historian James Blaut. Right up to the time of his death in 2000 he held that the world's central problem lay in its North-South inequities: tackling that justice issue would require a whole new view of global relationships. Commenting on Blaut's vision, Geoffrey Parker of the University of Birmingham points out that "an interesting attempt to achieve this cartographically is the Peters projection."[1]

Recent events may be lending credence to such convictions, including the momentous expansion of the G7 (or G8) nations to become the G20. Emerging powers such as China, India, and Argentina have suddenly become partners, not just spectators, in shaping the future of the world. They have an equal place at the table, with a full measure of responsibility. The world as visualized on Mercator-like maps is in process of being replaced.

James Wolfensohn, the Australian former head of the World Bank, also calls for a new perspective on the world. "Without dealing with [the] question of poverty, there can't be any peace." Yet the world is preoccupied with what he sees as parochial problems. Yes, in a striking departure from common understandings, he looks on issues like terrorism, Iraq, Afghanistan, strains in the Atlantic alliance, and budget deficits as minor – mere nitpicking! – compared with worldwide inequality. As a result of our focus on side issues, he asserts, the real problem – *grinding poverty and all that goes with it* – gets only "lip service." He highlights the figures: this is a world that manages to find $900 billion (U.S.) for the military every year, and $300 billion in farm subsidies in the industrial countries, but only $50 billion to $60 billion for global aid. "Absurd" is the label he applies before calling for widespread "missionary zeal" to deal with poverty and threats to the environment.[2]

In late September 2008, the media focused people's attention on the financial crisis that struck Wall Street. Day after day, month after month they reported – in far more detail than most people could absorb – what was happening in the United States Congress, in the financial markets, what might happen to international capital markets, to investments and pensions. Less than 5 miles away from "The Street," the United Nations had a very different agenda: the Millennium Development Goals. In contrast to a flood of information about stock prices, the story from the UN, with its call to commitment to overcome the scourges of disease, poverty, and hunger, was largely ignored.

The contrast reflects the disconnect between the "First" and the "Two-Thirds" worlds: the "rich" minority preoccupied with investments; the poor-world majority constantly concerned for food or a job or minimal health care or a bit of dignity. Does this mean we should turn away from issues of unemployment or corporate finance? No! They will always be foundational. But it is becoming clear that to ignore the well-being of the larger world is to fail to move toward the world we want. It is a sign of our global schizophrenia that "North" and "South" see and experience the priorities so differently.

When well-off people complain that the poor aren't doing enough to better themselves, famed economist

> **The world is fractured. That is serious, but not the worst news. More shocking still is the fact that so few are shocked by it.**

Jeffrey Sachs has an answer. Taking Africa as an example, he says its people are trapped by geographical handicaps: lack of navigable rivers, isolation, climate, and disease. To his list let us add another: trapped by widespread misperceptions of the importance of the South and its people, aided and abetted by the distorted maps in our minds. Weighed down with such heavy burdens, says Sachs, the director of the Earth Institute at Columbia University, neither the poor nor their nations can get out of the hole on their own.

Offering another prescription is the (then) director-general of the United Nations' nuclear watchdog agency, Mohamed ElBaradei. He points to – yes – weapons of mass destruction. Some warheads have been destroyed, he reminds us, but some 27,000 still stand in stockpiles. And who has these weapons of mass destruction? "More than 15 years after the end of the Cold War, it is incomprehensible to many that the major nuclear-weapons states operate with their arsenals on hair-trigger alert," he stated in accepting the 2005 Nobel Peace Prize. His clear point: When the present nuclear powers sharply cut their threatening stockpiles and redirect spending toward positive goals: combating poverty and dealing with organized crime, they will have made a "good start" toward a better life for all.

Why care about the world? First: 18 million human beings die of hunger and preventable illness every year, year after year after year. What other reason for action do we need if we believe that people are important? It would be gratifying if we could add that people and governments everywhere are responding heroically and the end of undeserved poverty is in sight. "Unfortunately, in the rich countries like ours, we really don't give a damn," is Jimmy Carter's stinging assessment. As a result, the world is entering a food crisis. Indicators point to prospects actually getting worse, not better. The distribution system denies even minimally adequate food to about one billion people; at the same time, many in the developed world consume more than their share and press a heavy carbon footprint on the environment. Can a world so out of whack long continue?

> "We who are on the right side of freedom's divide have an obligation to help those unlucky enough to be born on the wrong side of that divide."
>
> Condeleeza Rice, former U.S. Secretary of State.

Within our human nature there is a tendency – sometimes powerful, sometimes toned down – to reject appeals to high idealism in favor of calculating benefits to oneself. This is the ages-old "What's in it for me?" approach. Do-good organizations run into it; until they offer incentives, all their invitations to take the road to a better world will often draw disappointing response. Is this the ultimate bad news; does it really mean that the way to a world of fairness is blocked by ineradicable selfishness?

Marilynne Robinson, an American writer with a Pulitzer among her many honors, tackles the question head-on. She refers to people in the armed forces, then reminds us "there's a very strong chance they will not live to enjoy" what they are fighting for. To her list of fighters we might add a long list of martyrs for a cause and a host of social activists who work tirelessly for others. As Martin Luther King, Jr. put it, "I may not live to see the Promised Land, but..." That is, sacrifice – which nobody really wants – begins to make sense when we take the long-range rather than the short view, and the "we" approach rather than the "me" view.

The parallel between those who give of themselves and many a radical mapmaker is instructive. In refusing to accept the dominant worldview as final truth, they are both profoundly countercultural. "Another way to see the world" is a description aptly applied to the Peters; it can also be applied to the Dymaxion, Hobo-Dyer, What's Up? South!, and many other maps. In similar fashion, all who live - or even die - for the common good show us another way to "do life..." not in following the crowd, not in accumulat-

ing "stuff," not in self-aggrandizement, but in commitment to widespread hope for a more just world.

We may not live to see the results of our changed perspectives, but somebody will. As people in former generations, through their altruism, have made our 21st century lives better – in some ways at least – than theirs, so we have opportunities to do good for people we will never get to know.

Yet...

Suppose that all such appeals leave us unmoved, and we simply fail to act. How about factoring in our own benefits, here in the developed world? In reality, there is significant payoff. To the surprise of many, the North also benefits when people in the South experience dignity and fairness.

Noreena Hertz provides a powerful example in her book *The Debt Threat*. The government of Pakistan, she points out, has been so burdened by the interest it must pay on loans that it had to cut expenses somewhere. So where? In a pattern grown all too familiar, it slashed support for public education. For a time that seemed to work; there were, after all, alternative schools, so students were not thrown out on the streets. But some of those schools – who knows how many? – fostered attitudes of mistrust, even hatred of the West or the whole non-Muslim world. The record strongly points to one conclusion: those schools have incubated anger and terrorism.

The story of Hafiz Hanif suggests as much. In its issue of September 13, 2010, *Newsweek* points out that for years the world had no clear account from Al Qaeda insiders on their thought processes or education, but "Hanif's account provides that view." Here it is, in summary.

In his mid-teens, Hanif, an Afghan, lived with his parents in Karachi, on Pakistan's coast. He was a bright student, with good grades in math and a mastery of several languages. Though his parents never approved his dreams of joining the jihad, their influence couldn't match the persuasive power of those with whom he came in contact at school and on the street. Leaving his "regular" school, he enrolled with about 30 other students in an Al Qaeda training program. His father and mother desperately want him to come home, but he is determined to fight, to use the terrorist tactics he has learned, and even die for what he now believes.

If Hanif had benefited from, shall we say, "normal" schooling, his story might have taken a very different direction. Given its present trajectory, its final act may be disastrous both for him and those he manages to blow up.

Greg Mortenson, well known for his *Three Cups of Tea* and *Stones into Schools*, sometimes described as "knowing Afghanistan better than any other American," relates that

> ...in October, 1994, a group of about two hundred young men, many of whom had grown up in the squalid refugee camps around the city of Peshawar, joined forces to launch a new jihad. The vast majority of these men had studied in hard-line madrasas, or religious schools... where they had been indoctrinated with a virulent and radical brand of Islamist ideology. Calling themselves the Taliban, a Pashto word that means "student of Islam," they crossed the Pakistan border and swarmed into the Afghan truck-stop town of Spin Boldak with the aim of restoring righteousness and stability by uniting the country under the banner of a "true Islamic order."[3]

The conviction that underfunded schools and biased teachers in faraway places may be a worldwide problem is supported by retired Navy Admiral Dennis C. Blair. Early in 2009, while director of National Intelligence, he told Congress that unrest stemming from the global economic crisis had become the number one security threat to the country. Not subversives sneaking across our borders, not terror-ists blowing up bridges or power plants, but people blindly lashing out against their own hopelessness. So rather than pulling back from international commitments in times when money is scarce, Blair's plea is to start paying serious attention, if only for reasons of national security. That is, Americans concerned for homeland security cannot do better than to work for economic justice everywhere. Ban Ki-Moon, UN General Secretary, agrees. "We must act... not just because it is the right thing to do but because it is in all of our enlightened self-interest."[4]

> "We have lived by the assumption that what was good for us would be good for the world. We have been wrong. We must change our lives, so that it will be possible to live by the contrary assumption that what is good for the world will be good for us."
>
> Wendell Berry, farmer-priest and environmental activist

In another practical application, a strong case can be made for radically improving health delivery systems worldwide. Since infectious diseases don't respect those artificial lines on maps that we call national boundaries, an outbreak "there" can threaten lives "here."

In the final analysis, some of us cling to the idea that we can make a difference. It's hard to prove that statistically, but it's a conviction that won't go away. We care about the world – ought to care about it, want to care about it – simply because we can. Like Mount Everest, it's *there*; what more do we need? And in that caring we begin to experience what it is to be human. Not just to be descended from some common ancestor, but expressing our solidarity with the whole of humankind – that is what marks us as members of the human race.

How to Care about the World?

Deciding to care – *really* care – about the world doesn't magically make everything clear. Chief among the sticky issues still to work on is how to do the caring.

> Predicting the future – which we all keep trying to do – can be very tricky. But here's the thing: the best way to predict the future is to *shape* the future.

In Chapter 8 we explored major outlines for the kind of world we want. Here we need to look, though briefly, at how we map the route. Two major approaches stand out: push ahead through exercising power (dominance and control) or seek to go forward by way of justice (fairness or equality). The tension between these two is evident all through Western culture: in the political arena, in faith/value systems, in funding for research and development, in social patterns. We have spoken about the chasm separating the global North from South or the affluent from the underprivileged; let me now add that the split between the advocates of power and the advocates of justice may be just as relevant to the future. The approach through power has its associations with military strength, terrorism/counterterrorism, imperialism, the weak serving the strong. It also connects, I hold, with maps that portray one part of the world dominating the rest of the planet.

The approach through justice, on the other hand, is aligned with emphasizing freedom from fear and want, the Earth sustaining all its people with adequate food, safe drinking water, decent public health and fair treatment. The association with equal-area maps will, I hope, be clear by this point in our conversation.

If the symbol of one approach is a clenched fist, the proper image of the other is an open hand – ready to work with the needy or to shake a stranger's hand. If the one is experiential (meaning that the world has long and depressing experience with it), the other is experimental (we are still struggling with how it could work). One approach is seen in pax romana (Rome's ability to silence all opposition, imposing what the Empire labelled peace) or some modern variant: the powerful applying pressure to make sure the "servants" stay in line. The other takes on reality through that portion of foreign aid that benefits the people more than powerful despots, and those voluntary organizations that work on behalf of refugees or for global health and education.

The contest continues, and the choice is real. Exactly how it will all work out one can only guess, but I am convinced that maps will play a major role in shaping the outcome. To that question we now turn.

Why Care about Maps?

Of all the reasons we might identify, let's focus on three.

For the Sake of Accuracy

If it seems obvious that maps should be accurate, let's be clear: some maps falsify. Consider examples.

A world map published in Castro's Cuba seems like what you would expect – in most respects. Africa, Asia, Europe look "normal." Some things, however, startle: Cuba, close to the visual center, leaps into enlarged prominence. Rays, as if from the sun, fan out from it – to suggest Cuba enlightening other nations? As for the USA, too bad - it's cut to a fraction of its proper size. A false perspective? Certainly. Yet it is just one example out of many of a mapmaker's agenda taking over. Like many maps, it puts a spin on reality.

Fig. 11-1 Cuba-centered map
Source: Author's collection

Map publishers have been known to insert a feature – say a street or a stream or a bend in a road – that doesn't exist, or to deliberately change the height of a mountain; it's one way to trap copyright violators. Whether for that reason or as a humorous prank, a mapping official secretly added two fictitious "towns" to Michigan's official 1978-79 road map. Locating them over the border in northern Ohio, he named them Goblu and Beatosu. In football, *Go Blue* was the University of Michigan's rallying cry; Beat OSU (Ohio State University) was their hope. Maybe he set a trap... he certainly snuck in a couple of plugs for his favored team.

In the Nazi era, German maps deliberately enlarged the size of the German Reich. Their propagandists clearly understood a reality that many even today have trouble crediting: that maps carry power. That is, their message burrows into the psyche, and from there it spills over into attitudes and action. In effect, they had laid hold of a basic fact: *maps are all about climate change*. That is, Nazis used distorted maps to change the climate of German public opinion. Today the challenge is to see maps as tools for a more accurate, more realistic view of the world, for the benefit not of one nation or one group but of the whole world.

At another level, two noted oceanographers contrast the differing perceptions about ocean temperatures – and consequently of global climate change – resulting from plotting temperatures on a Mercator image and an equal-area projection. "There are many ways to project a spherical surface into a plane," assert Gunther Krause and Matthias Tomczak, "some are more useful for a given application than others... [A]rea is important because the coupling between the ocean and the atmosphere occurs at the sea surface, and most exchange processes are proportional to the area of contact." Their conclusion? That the Mercator and certain other maps are seriously misleading, whereas "The Peters... will often be the best choice."[5] Never underestimate such an insight as trivial. A conclusion like this can affect quality of life for you and me – indeed, for people everywhere. Like these oceanographers, climate scientists are seeking

out the most appropriate maps for their purposes. Based on best models, they now predict – in 2012 – severe droughts for the next 20 years. These will hit many areas of the world, including such major food-producing areas as the American heartland.

To raise the question of accuracy is also to ask what gets included and what gets excluded. Maps of "local attractions," commonly used by visitors, usually fail to reveal how a place qualifies to get on the map. In some cases at least, the criterion is first and foremost financial: organizations that take out advertising or otherwise pay get on, while other sites equally worthy are bypassed. So unsuspecting tourists are fed skewed information. More specifically, J. B. Harley, British geographer, professor at the University of Wisconsin-Madison and co-author of the renowned *History of Cartography*, points out that the United States Geological Survey, the trusted source for highly detailed, official data about the geography and topography of the country, glossed over nuclear waste dumps as if they weren't even there,[6] all the while providing data on places of much less social concern.

Think again about Fig. 1-4, the image of Iraq set before the Energy Task Force. At first it seems accurate and acceptable: shape, scale, position, and other map qualities are preserved. Still, the nagging question remains: by making Iraq look more like a vast oil reservoir than a place where people live and work and eat and love, did it skew reality?

Easy answers tend to be wrong answers, of course. But the question is worth asking, and working on answers may open up insights not only into how maps work but how political decision makers deal with their world – indeed, how all of us deal with the world. If a map, by the way it selectively presents data, can nudge otherwise intelligent persons toward a particular perspective – whether helpful or tragic – where are the checks and balances in the system? Are we ever safe from mapping manipulation? How can we learn to decipher the messages embodied in maps?

In sharp contrast to all examples of deception (intentional or not), really caring about maps means that questions of perspective, accuracy, and reliability always belong front and center.

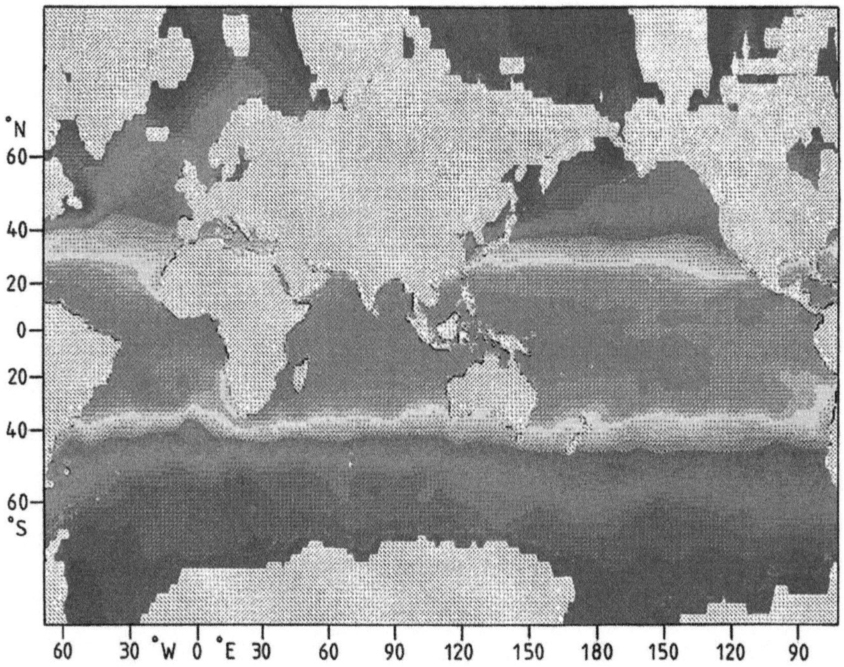

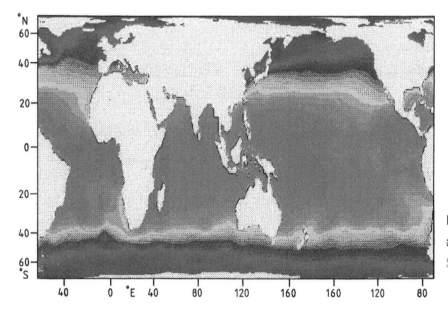

Fig. 11-2A Ocean temperatures shown on a Mercator Projection

Fig. 11-2B Ocean temperatures shown on a Peters Projection
Source: www.tos.org

For Our Own Sake

We, the people of all nations, deserve the fullest, most functional map and view of our global home that it is possible to achieve. In practical terms that means not one map but many. Through internalizing many ways to see the world we may even develop openness to other people's points of view and greater self-awareness. For example, a map that sets South at the top – on any projection you choose – may stretch understanding. And certainly we hold that equal-area maps belong in every world citizen's collection.

Civilizations do not drop down, fully formed, from the sky; they develop over time out of a people's sense of who they are. Self-understanding gives rise to national anthems, systems of governing, attitudes toward other groups. Colonial conquest, the Marshall Plan, or intervention in the life of another nation arise out of our worldviews, which in turn are conditioned by our maps.

Since both desirable (moving toward the world we want) and undesirable (pushing us into trouble) events stem in part from our map-based perspectives on the world, does it not follow logically that we ought to care deeply about our maps? When we get them wrong, everybody suffers; when we get them right, the common good stands to gain.

For the Sake of the Cheated People

As the Prime Minister of Malaysia was leaving office in 2003, he spoke frankly. To outsiders, his speech sounds like variations on the theme of being put down. "I will not enumerate the instances of our humiliation," said Mahathir Mohamad. "We are all Muslims. We are all oppressed. We are all being humiliated... Today we, the whole Muslim [community] are treated with contempt and dishonor." Though he had led his country for two decades during which its economy showed remarkable success, he went on to speak of "hopelessness" and a feeling among many Muslims that they can do "nothing right."

> "Six and a half billion humans must make a choice to change course, to turn to life as our defining value and to partnership as our model for relations with one another and the planet. [We must] free ourselves from Empire's cultural trance by changing the stories by which we define our possibilities and responsibilities."
>
> David Korten,[7] self-styled conservative, leader in the movement for a new economic and social order.

Korten may be right: changing the stories we live by will make a difference. But since, as the saying goes, a picture is worth a thousand words, I assert that it also makes sense to change the maps we live with!

> "Look at the world around you. It may seem like an immovable, implacable place. It is not. With the slightest push – in just the right place – it can be tipped."
>
> Malcolm Gladwell, *The Tipping Point*

But you don't have to be Muslim to be put down. "Do not degrade the people of the tropics," was Arno Peters' advice to mapmakers. "They're having a tough enough time as it is." The goal, then, is first to understand the world as fully as possible – which requires appropriate and provocative maps – then to reshape that world, which means the world's peoples acting, separately and together, to make real the world we want. To care about maps is only one step – yet an absolutely necessary one – in the challenge of redesigning how we experience the world.

In an example of great insight, a group of thinker-activists once – it was July 4, 1776 – called for a "decent respect for the opinions of mankind." Out of that perspective on human affairs came the breathtaking experiment that we call the United States of America. Today the question is whether the developed world, now led by the United States, is adequately demonstrating such respect for the opinions of the rest of humankind. Ironically, much of the Third World now sees the privileged world the way 18th century American colonists saw Britain – as uncaring and unresponsive. Why should we be surprised if, like American colonists, they take matters into their own hands? What would "a decent respect" for their opinions mean?

Why Care, Anyway?

We opened this chapter with an assertion that cataclysmic events may drain so much energy from people that they reduce their caring capacity. This may be such a moment in history. Even as the numbers of the very needy are increasing, the readiness of the privileged to share is under serious pressure. To be specific: in June 2009 the United Nations' Food and Agricultural Organization reported that over 1.2 billion people go hungry day after day; this is the highest number in history. At the same time governments at every level face shrinking revenue streams, forcing them to curtail programs of assistance to the needy at home and abroad. Average citizens may also experience compassion fatigue: they simply cannot keep up with the mounting appeals. The spirit may be willing, but the wallet is anemic.

Perhaps, instead, tough times should fire them up, redoubling their good work. Can we expect that to happen? Not if you're a realist! Far from strengthening assistance to the world's neediest, popular opinion often sees aid as the first place to cut when budgets need balancing. Americans are sometimes surprised to learn that foreign aid amounts to just one percent of the nation's budget, including military aid, so trying to solve our financial troubles that way will hardly show on the scale. But reducing or eliminating such foreign aid will throw many into poverty and an early death and handicap the nation's global policy initiatives. And, equally important, it will reduce our sense of connectedness with all of the world's people.

Into this heated debate, throw the rising public desire of many for compassion, fairness, and hope.

I see this in the action of the people of First Baptist Church of Orlando who, after a segment on TV's *60 Minutes* revealing the reality of homelessness in Central Florida, raised, over a period of just two weeks, $5.6 million beyond their regular contributions, to help. This action was further boosted by the show's wide audience, which sent in an additional million dollars plus. I see this will to change expressed in the Arab Spring, the Tea Party phenomenon, and the Occupy movement that began on Wall Street in 2011 and soon spread across the USA and beyond. In such disparate movements there is hope. Combining high idealism and a realistic view of our global community can only bring positive results.

One who understands that well is Maria de Lourdes Pintasilgo, President of the Independent Commission on Population and Quality of Life. She points to the urgency as well as the need to care: "The three decades ahead may prove to be the most critical in history."[8] The moral challenge of our time, she asserts, is to expand our caring. We can do this by doing away with what she calls forced poverty, by assuring people's economic, social, and political rights, and by lending support to raising the quality of life for all.

Wayne Gretzky, sometimes called the greatest hockey player of all time, puts it in simpler terms: "You miss 100 percent of the shots you never take." Scoring the winning goal is never guaranteed; what we can do is give it our best shot. At least then there's hope.

CHAPTER 12

Let the Conversation Continue...

"Endings to be useful must be inconclusive."

Samuel Delany. Award-winning science fiction author.

The world is a lot like the weather: it'll change. So will our maps. And our way of understanding maps. And our – well, what the Germans call *Weltanschauung* – the worldview that shapes the way we live.

Which means, among other things, that this book won't be the end of the discovery process. Of course, it was never intended to be the last word; it's time now to confront that fact quite openly. Some people will find these pages helpful, expanding their perspective; others will quarrel with one assertion or another or with some of the persons or organizations called as witnesses. Some may object to the underlying premise: that our maps and our relationships with the world are intimately related. No problem: the goal is not to exact agreement but to stimulate thinking.

Indeed, the best conversations always continue. The conversation opened up in this book now needs to move forward on three levels:

- The meaning of maps: what they are, what they do, how they work.
- The special contribution of equal-area maps.
- The world's problems and possibilities, with special attention to how developing new perspectives, particularly evidenced in maps, can contribute to a positive future.

Continuing the Conversation: Maps

A hundred years ago, if you wanted to be a mapmaker you learned how to measure and draw, then went into the world to observe and record.

Two or three centuries before that you would still have surveyed and recorded, but you might have journeyed in a sailing ship to distant coasts, mapping shorelines and navigable rivers. You might have filled in the blank spaces with fantastic creatures.

Today you are more likely to look at a computer screen while you click on data points. It hardly matters whether you've mastered how to use a sextant or the principle of triangulation or surveyed the scene with your own eyes; with the aid of good software anyone can learn to customize a map. What we are experiencing in our time is a dramatic democratizing of the mapmaker's art. More and more people are showing at least a basic interest and even some degree of skill in the field.

That is progress. If Google Earth is to be believed, 80 percent of the world's data has a geographic component. In a world like that, what's not to love about people learning to appreciate maps and how maps connect us with the world?

Maps, in fact, have an expanding usefulness. The entire mapping process has unique, practical, and increasingly important contributions to make in daily life.

That said, progress is never simple and straightforward. For example,

Will GPS Kill Geo-Literacy?

The hand-lettered sign on the gas pump was clear: Please pay inside before pumping.

I did. I said I wanted to fill the tank. Regular.

"Nope," came the reply. "Tell me how much you want."

"It'll take ten gallons."

That was not what she needed. "Give it to me in dollars and cents."

"Well, at $2.95[9] a gallon, ten gallons will be $29.59, right?"

The attendant looked at me as if I had just breezed in from an unknown planet. After a moment of silence she called out to another worker, "Tom, this guy wants ten gallons. What does that come to? I don't have my calculator with me."

Some educators are convinced that over-reliance on calculators results in people not being math literate. Perhaps that was the case with this attendant: she had a high school diploma, yet evidently did not know that to multiply by ten you merely move the decimal point – one of the simplest mathematical exercises ever devised. She was numeracy-challenged.

Is there a similar danger when we depend on GPS devices to tell us how to go where we want to go? Can the machines that supply the answers atrophy that part of the brain that formerly figured out the answers? If we don't exercise our spatial consciousness, will we eventually lose it?

The question is in the earliest stages of being asked, so no one knows for sure. But what would it be like to live in a world where even the simplest reference to mapping or to geographical understanding had to be referred to a machine because we had lost the ability to intuit the answer in our brain?

It's far too early to panic. But it's not too early to ask the question.

Through Maps to Better Public Health

A good example of maps contributing to the common life comes out of 19th-century London. Actually, there are two stories: one overlaid with legend, the other more complex and accurate.

The popular version, in outline, opens as thousands are dying of cholera. Marking up a local map, a physician named John Snow keeps track of where the deaths occur. Struck by the clustering of fatalities in the area served by a particular water pump – on Broad Street – he presents his findings to the Cholera Inquiry Committee in 1855. From the data, he develops the hypothesis that cholera results from contaminated water. His peers, however, are unimpressed; their minds have locked in on the idea that

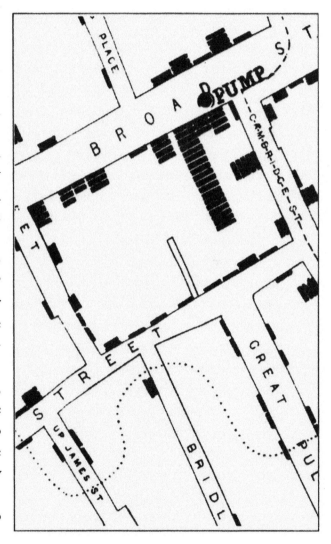

Fig. 12-1 John Snow supposedly used this map to creatively link statistics and geography: the numbers dying of cholera and the location: where the victims lived and where they got their water. Notice how many bars, representing one or more deaths in a household, are grouped around the Broad Street pump. Mapping led to a medical advance.

Source: Brody H. et al. *The Lancet* 356(9223), 64-68, 2000.

the disease is airborne. The breakthrough comes when Snow persuades decision makers to take the handle off the Broad Street pump. Deaths decline dramatically. So Snow makes his point; uncounted lives are saved. And he has effectively pioneered the medical use of mapping. It is not irrelevant to add that in 2003 Snow was declared "the greatest physician who ever lived" based on a poll of British medical personnel. (Hippocrates, the iconic founder of medicine, came in second.) The connection between his status as a medical mapper, holding the man in veneration (not too strong a word), and voting him number 1 in his profession seems clear.

Recent research reveals a more layered, if less reverential, story. Tom Koch, a geographer at the University of British Columbia, has made it clear that Snow did not single-handedly solve the cholera puzzle. He was not the first to use maps in medical sleuthing; neither was he totally clueless about the cause of cholera until his map made everything perfectly plain. Perhaps most pointedly, the maps even many professionals still assume to be Snow's may not be his at all. Let's summarize.[1]

- Snow utilized existing maps, 1854 and 1855, marking cholera deaths by location.
- E. W. Gilbert adapted Snow's work, passing off the result as "Snow's map." His interpretation appeared as "Pioneer Maps of Health and Disease in England" in 1958.
- Edward Tufte in turn revised Gilbert's adaptation for his *The Visual Display of Quantitative Information*, 1983.
- Mark Monmonier further modified Tufte's result for his *How to Lie with Maps*, 1991.[2]
- The U.S. Centers for Disease Control (CDC) utilized this in 2001, so lending the prestige of their office to a flawed understanding.

As a result, Koch points out, "the map itself is twisted, turned, and truncated – violated by each mapmaker's mindset and point of view." Would Snow recognize the maps and the medical accomplishments attributed to him? Doubtful.

Koch further points out that the line of reasoning was not as straight as may be supposed. Not "my maps *prove* that the pump is the problem," and "the pump in turn proves that cholera must be transmitted by water."

Rather, the marked-up maps laid open a chain of events like this:

Living closer to the Broad Street pump →higher probability of people drinking water from it →heightened death rate. That is, the maps did not serve as decisive tools of analysis so much as pointers to a possible cause-effect connection.

In both accounts one thing is sure: maps mattered. They contributed to controlling the outbreak of cholera in London. If you and I go through the day without fear of cholera, we benefit from the historic advance that clusters around Snow and his colleagues. There was a synergy between using maps and medical progress.

If Snow became a pioneer in the application of mapping to questions of public health, others have developed it further. In the textbook *Endemic Areas of Tropical Infections*, an international publisher uses more than 100 maps to show exactly where certain infectious diseases are prevalent. Significantly, author Dieter Stuerchler demonstrates a good grasp of how maps work: he uses the Peters whenever he needs a world map, thus taking advantage of its equal-area quality, and a more shape-accurate equal area map when he wants a smaller region, say Africa.

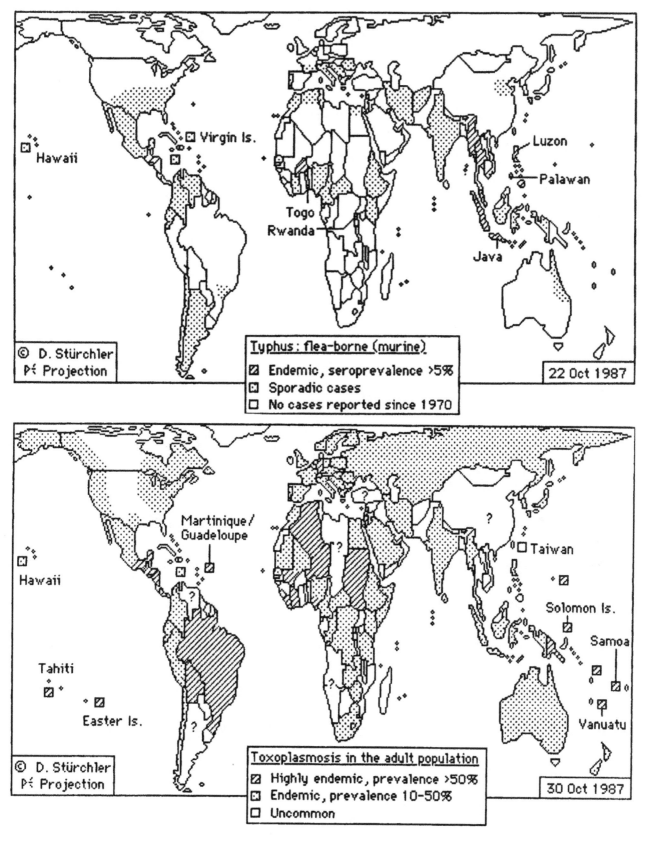

Fig. 12-2 A and **Fig. 12-2 B** Two examples of mapping the prevalence of disease: A is flea-borne Typhus; B, Toxoplasmosis in adult populations. The choice of an area-accurate map provides for comparability across regions that might otherwise be lost.

Source: © 1988, Dr. Dieter Stuerchler. From *Endemic Areas of Tropical Infections*, published by Hans Huber Publications, Bern. Used by permission of the author.

Can the Right Maps Lower Crime Rates?

Crime – think of it as another form of disease, a pathology of how we live in community – is also being mapped in efforts to control it. Using sophisticated techniques developed in University College, London, crime prevention crews conclude that people and property within 400 meters of, say, a breaking and entering offense are at elevated risk for up to two months. So the crews draw up maps of "prospective hot spots." Some police forces claim 30 percent stepped-up effectiveness compared with other approaches.[3] The Los Angeles Police Department has been effectively using this approach to allocate scarce human resources in the most effective way.

In 2008 and 2009, the *Toronto Star* released attention-getting analyses that some predict will at least bring about a rethinking of public policy on prisons, possibly even lead to a new approach for dealing with criminals. Mapping served as the essential tool. The entire Greater Toronto Area – Canada's largest – was mapped. Incarceration costs for each postal code were overlaid. These were then compared with such social indicators as education level, unemployment, and percentage of households headed by a single female in those same areas. Those results were then set alongside available data on both the costs and benefits of intervention: providing preschool programs, recreational facilities and job training, for example.

The bottom line? Taxpayers are spending more to keep the lock-up system running than they are to help the most vulnerable live useful lives. Some analysts even conclude that prison is largely ineffective at reducing crime. Prevention, on the other hand, pays off. Among observers with a professional vantage point is Patrick Lesage, former chief justice of the Superior Court of Ontario. "The root causes of crime and violence aren't resolved by putting people in jail," he comments.

Both mapping as an essential tool and the conclusions of the Toronto study are paralleled by other, independent researchers. The Justice Mapping Center, based in Brooklyn, has broad experience in using computer mapping applied to issues such as criminal justice. Working with government agencies and other concerned groups, it has enabled decision-makers in Gateshead, UK; Wichita, Kansas; the State of New Jersey and other places, to see crime questions in a new perspective, and hence deal with them more effectively. Sending people to prison begins to look like a simplistic nonsolution rather than a rational response to the problem, according to some experts in the field. Thus Eric Cadora, a cofounder of the Justice Mapping Center, says, "I think of incarceration as a lazy response to poverty. It's a way of not dealing with the issue."[4]

A Mapping Approach to Climate Change

Can we see not only cholera and crime but also global warming as a malady, one from which all people as well as the nonhuman world suffers? Do maps matter in such an issue?

At the very least, as climate changes, it is clear that maps will have to be redrawn. As sea levels rise as a consequence of melting ice caps, beaches will recede, small islands simply disappear, and port cities such as New York, Vancouver, Amsterdam, Shanghai, and Monrovia will have to deal with water levels predicted to go higher, from one to seven meters (3 to 23 feet) over the next several decades. "Waterfront property," traditionally in high demand, may become something to avoid. Given present realities of how resources are distributed, the impact of rising sea levels will be felt far more severely among the poor. As ocean temperatures rise, typhoons and hurricanes will become a greater threat, further eroding shoreline contours. Extreme weather may become the norm, not the exception. The years 2011 and 2012 stand out for the devastating weather problems that hit wide swaths of the planet.

Climatologists warn that, paradoxically, inland waterways will shrink for lack of water. In some places, the productivity of land will decline as agricultural soil turns to desert. At a bare minimum, then, mapmakers will keep busy redrawing shipping charts, commercial maps, political maps, property line maps, topographic maps.

But maps themselves are not the heart of the problem; more serious is the human reality to which they point. Lives are being disrupted – or ended – and will continue to be.

Maps: A Tool for Social Action

Is there more? Yes! Maps not only set forth present reality and expected futures; they can alert us to alternate scenarios, giving impetus to present action. In central California, the Big Creek Lumber Company wanted to log on land owned by the San Jose Water Company. The lumber company stated that the area affected measured 2,002 acres, which under existing regulations gave them legal harvesting rights. In effect, they had a green light. They notified local residents of their plan, including a map.

Large numbers of people either assumed the proposal met all requirements, or ignored it. Many said they couldn't understand the map, so disregarded it. They turned their attention to more immediate concerns: washing the car, shopping, firing up the barbeque.

Nevertheless, a few formed Neighbors Against Irresponsible Logging (NAIL) to oppose what they saw happening. One of their worries was that water quality would be affected, the area in question being part of the watershed that supplies a quarter of the drinking water in Silicon Valley.

With the help of volunteers including Rebecca Moore and Adelia Barber, a Ph.D. student at the University of California Santa Cruz, the group used such tools as Google Earth to do a detailed analysis of the logging proposal. One result: they mapped the affected area at 2,733 acres – which meant it violated the law. Their map also showed that the logging area would come close to a preschool and day care center – the buffer would be less than 300 yards wide. At that point their mapping took on two new dimensions; it became a political argument against logging and it delivered an emotional wallop. Not at the expense of precision, however. Since there are "many ways to map," they pointed out, they made sure their process followed the best, most reliable techniques. To counter, Big Creek Lumber prepared a modified version of its map, but NAIL's persuasive mapping and careful calculations were upheld by the California Department of Forestry.

We're all familiar with protests involving boycotts, barricades, marches and demonstrations. We also know the attitude of "You can't fight City Hall," or that a handful of people can't stop progress or oppose powerful companies. NAIL's action contradicts that conventional wisdom. "We were sitting at our computers and slaving away – and it worked," commented Barber. Their secret weapon? Good mapping.

This effort – by amateurs in mapping! – gained support from environmental activist former Vice- President Al Gore. More than that, it has inspired others, including the Sierra Club of Minnesota as well as citizen groups as far away as Australia. The value of Google Earth and of mapping as an effective tool has been made clear – and may safely be predicted to become more widespread.

What if people all over the world threatened by, well, you name it – logging, commercial development, polluting factories, hydrofracking chemicals in drinking water, say – what if they could all see maps as tools of analysis and action? What if residents of the Florida Keys and power brokers on Wall Street could see a believable map that showed their familiar venues under water? How about a map to make clear where the hazardous electronic wastes of the developed world get dumped – in parts of the world where rent is cheap and human life is devalued? What if we could visualize the human devastation from rising sea levels to people eking out a living among the waterfront shacks in Dhaka, Bangladesh? Can maps make a difference? How can they not? Bring together maps, available technology, human creativity, and people's willingness to take a stand and you've got a powerful recipe for changing events.

Continuing the Conversation:
Equal-Area Maps

Arno Peters, we have said, had a clear intent when he created his map. He believed that a fairer map would move us toward a fairer world.

Obviously, we have not achieved that fairer world, though the longing for it is broad and persistent. Similarly, equal-area maps are not the standard image in the media or in people's minds. Maps that were a source of pride in our grandparents' era still skew our perceptions. There has been progress, but the hangover from our long binge on old map images continues, and there is a long way to go.

Progress: that is the first reality in the continuing story. The buzz that the Peters map has developed – its unprecedented rise to prominence and the controversy it has stirred up – coupled with our heightened awareness of global realities now clamoring for attention, these all work together to ensure that the story will go on. The Peters is arguably the pre-eminent map story of the late 20th century. Some believe its full impact will be part of the history of the 21st.

To turn from the general to the specific, consider the rain forest of the Amazon basin. Sometimes called the "lungs of the world," this vast expanse cleanses much of our global air, taking in carbon dioxide and releasing life-supporting oxygen. It is said to be the largest remaining stand of timber in the world. But only on an area-reliable map does its extent come clear. If that extent is under threat – as many experts are warning – air quality is in trouble globally, not just locally. Area-accurate maps speak to our need to understand our world, those resources on which our life depends, and the absolute necessity of acting as good stewards of the Earth.

Note that we use the term *conversation* to describe this book: our intention is that you find these pages an exploration of some important questions. Now let's apply the term another way: a map, just as this book does, invites conversation. Roll that around in your mind: *Every map is an invitation to a conversation.* It elicits dialogue between you, the viewer, and the physical product we call the map. Or, if you will, between you and the mapmaker's agenda. In the case of the Peters and the Hobo-Dyer and Goode's Homolosine, it becomes an opportunity to participate in a new perspective on the world – to take seriously certain realities of life on the planet. As you enter into dialogue with the visual you gain a new vision; that, in turn, can motivate action. It does happen! That is precisely why, in a book that explores maps as tools for understanding the world, we have opened up such issues as poverty and hunger, war and peace, resentment and hope, condescension and respect. Questions of liberty, equality, the pursuit of happiness, ethics, public policy, foreign affairs, intercultural relations, economic justice, the environment, and how we live together on planet Earth, our home – these are never far from any dialogue about today's provocative maps.

To participate in such a dialogue is to engage in reframing the questions that define our common life. This book presents the question: Can we see – *really* see – the vital connection between the maps we use and carry in our heads, and the way we deal with the world?

Continuing the Conversation: What Lives Are Worth Saving?

The question is well-posed by Bill Gates, founder of Microsoft. Gates was shocked to learn, some years ago, that in the developing world about a half million children die each year from rotavirus, which brings on severe diarrhea. "Why have I never even heard of a disease that kills half a million children every year?" he asked himself.

There was more to come: Gates learned that millions of children die of diseases that pose no threat in developed countries. He and his wife, Melinda, decided to take action. Their foundation – said to be the largest in the world – focuses on the medical needs of people on the other side of the world's privilege gap. Speaking to the World Health Organization in Geneva in 2005, they summed it up: they "couldn't escape the brutal conclusion that – in our world today – some lives are seen as worth saving and others are not."

The point is worth pondering. Are Bill and Melinda Gates right? If people attach high value to some lives and minimal value to others, why is this so? Pursue the issue further: Are some people not only perceived as more valuable but *are actually* more worthy of health and opportunity than others? Are some nations somehow entitled to hold on to special privilege while others are denied equal opportunity? Is exploitation of the weak by the strong – the master/slave relationship that we thought was gone for good – now reappearing in the way whole nations deal with one another? To what degree should equal treatment for all be a matter of principle? And how is that to be reflected in world maps?

Continuing the Conversation: Global Issues

In late September 2009 the G20 – nineteen nations plus the European Union – met in what has come to be known as the Pittsburgh Summit. The setting was significant: delegates sat in a circle around a map of the world that, in the words of a spokesperson, imaged "unity and inclusiveness as well as diversity." National borders were deliberately omitted. Given its massive size and centrality, it's hardly surprising to learn that the map drew lots of oohs and aahs. Images of the map were sent around the world with the final communique.

"The map symbolizes President Obama's message that today's economic and environmental changes know no borders, and will require global solutions. The consequences of failure (he and his team have said) are global in scope."[5]

For the record, the map seems to have been based on the Miller Cylindrical with some modifications by the designer (artists do sometimes take liberties with strict facts) set within a circle. The point here is not whether the choice of projection was the best; rather, that summit planners thought globally and to that end set a global image before the delegates in a way that would keep the good of all – not just of any one segment – in their focus. Will this image become "the map shot 'round the world" – not only because it formed an integral part of the Summit report, but because it heralds a new era in international cooperation? So the conversation about maps and global issues continues and can make a difference.

A few paragraphs ago I referred to some of the world's most pressing issues. But underlying them all is one persistent issue: how we as humans relate to one another, to the world, to those values set out in the UN Declaration of Human Rights and regularly given lip service. Is fair treatment for all even possible? And if so, what would that look like?

Far from being spun out of airy idealism, such questions are intensely practical. General Roméo Dallaire says that at the Canadian Forces Peace Support Training Centre instructors use a chart to make plain the nature of the world. The chart looks much like this:

STATE OF THE VILLAGE REPORT

If the world were a village of only 100 people, there would be:

60 Asians,
14 Africans,
12 Europeans,
8 people from Central and South America, Mexico and the Caribbean,
5 from the USA and Canada, and
1 person from Australia or New Zealand.

The people of the village would have considerable difficulty communicating:

14 people would speak Mandarin,
8 people would speak Hindi/Urdu,
8 English,
7 Spanish,
4 Russian,
4 Arabic.
This list accounts for less than half the villagers. The others speak (in descending order of frequency) Bengali, Portuguese, Indonesian, Japanese, German, French, and 200 other languages.

In the village there would be:

33 Christians,
22 Moslems,
15 Hindus,
14 Nonreligious, Agnostics, or Atheists,
6 Buddhists,
10 all other religions.

In this 100-person community:

80 would live in substandard housing.
67 adults live in the village; and half of them would be illiterate.
50 would suffer from malnutrition.
33 would not have access to clean, safe drinking water.
24 people would not have any electricity.
Of the 76 that do have electricity, most would use it only for light at night.
In the village would be 42 radios, 24 televisions, 14 telephones, and 7 computers
   (some villagers own more than one of each).
7 people would own an automobile (some of them more than one).
5 people would possess 32% of the entire village's wealth, and these would all be from the USA.
The poorest one-third of the people would receive only 3% of the income of the village.

The following is also something to ponder...

If you woke up this morning healthy ... you are more blessed than the million who will not survive this week.
If you have never experienced the danger of battle, the fear and loneliness of imprisonment, the agony of torture,
   or the pain of starvation ... you are better off than 500 million people in the world.
If you have food in the refrigerator, clothes on your back, a roof overhead and a place to sleep ...
   you are more comfortable than 75% of the people in this world.
If you have money in the bank, in your wallet, and spare change in a dish someplace ...
   you are among the top 8% of the world's wealthy.
If you can read this, you are more blessed than over two billion people in the world who cannot read at all.

When one considers our world from such a compressed perspective, it becomes both evident and vital that education, acceptance and compassion are essential for the progress of humankind.

This material is copyright-free. You may reproduce it.
Original version by Donella H. Meadows

More info at **www.odt.org/pop.htm**

Fig. 12-3 State of the Village Report.

Source: Donella Meadows. 2005 data provided courtesy of Sustainability Institute and produced as a public
service (copyright free) by ODTmaps. This document is a PDF online at: www.odt.org/Pictures/popvillage.pdf [6]

Clearly, most of the world's residents – even when we celebrate the fact that millions in such places as Brazil, China, and India are moving out of poverty – live in conditions sharply different from those considered normal in the Northern world. That has consequences, points out Roméo Dallaire:

> ...many signs point to the fact that the youth of the Third World will no longer tolerate living in circumstances that give them no hope for the future. From the young boys I met in the demobilization camps in Sierra Leone to the suicide bombers of Palestine and Chechnya... we can no longer afford to ignore them.[7]

General Dallaire, currently a Canadian Senator, goes on to speak of predictable results. "This lack of hope in the future is the root cause of rage. If we cannot provide hope for the... masses of the world, then the future will be nothing but a repeat of Rwanda, Sierra Leone, the Congo and September 11."[8]

The concern exists: it shows in the statements and actions of many to whom we have referred in this book; it lives also in the hope and will of people across cultures, nations, faiths. All these are part of the continuing conversation, without which the world seems destined to continue in a downward spiral.

Dallaire, like Bill and Melinda Gates, keeps coming back to a single question: "Are we all human, or are some more human than others?" If we answer that we are, indeed, equally human, does it not follow that we should choose maps that reflect that equality? If we say we are not yet equally human but that would be a goal to strive for, does it not make sense to employ maps that visualize that future and that draw us toward a world of fairness?

If we are honest, we must confess that we, the people of the world, do not act as if all deserve equal treatment. A news reporter was asked about the world's seeming unconcern while hundreds of thousands die in Darfur, compared with the response she had seen to wars and natural disasters in other places. "Over here, they're just dead Africans,"[9] she replied. We have not taken that long journey to a new mindset that makes for fairness to all. Most of us haven't even, yet, taken the first step: fully utilizing and accepting a fair-to-all-nations world map!

Is the situation hopeless? By no means. Dallaire closes his book with a can-do attitude: "*Peux ce que veux. Allons-y*" – literally, "We can do what we want to do. Let's get on with it."

Stephen Lewis, whose experience includes service as a politician, UN diplomat, professor, AIDS activist and humanitarian, aptly titled a series of lectures, "Race against Time."[10] "Every minute delayed is another life lost." While his focus was clearly on HIV/AIDS across Africa; and while he made an impassioned and moving plea for the world to respond urgently to that human crisis, the present global need encompasses more than one health issue. We are all in a race against time. Can the world be healed before it moves further and ever faster into tragedy? By our action – or inaction – we are supplying the answer. No one has the time to do it all. The question becomes "how deep is our commitment to fixing the world?"

Where Dallaire says we can do it if we will, where Lewis sounds the call to enlist now, others point out that even when the cause seems hopeless, prompt, strong response can make an impact. One of this last group is William Schultz, who formerly headed Amnesty International, a worldwide movement to free political prisoners.

> About a year ago, a church in the little town of Bedford, Massachusetts (I know it well because I was a minister there in the 1970s) hung a banner over its front door protesting the march to war in Iraq. The church is located in a prominent place in town – right in the town square – and Bedford is the home of Hansford Air Force Base so the combination proved quite explosive. Day after day military families and their supporters demonstrated in front of the church, objecting to what they regarded as an unpatriotic banner in the heart of their small town.
>
> War came of course and people started dying and one day last fall two military police appeared at the door of a Bedford family.

That afternoon the minister received a call from a stranger. "May I come see you?" the stranger asked. "Of course," the minister replied, and when the stranger arrived, this is what he said, "This morning I learned that my son was killed in Iraq. Last spring I was one of those who demonstrated against your church and its banner. But today I realize I was wrong. You are the only ones in town who tried to prevent this madness. I'm here to ask your forgiveness and one thing more: I'm here to ask you to do the memorial service at Arlington National Cemetery for my son.[11]

Embedded in the story is good news: Anger can give way to acceptance. Resentment can turn to respect. The "no" that seems so final can grow into "yes." New perspectives can make a difference. Let the conversation continue!

That ongoing conversation will gain practical application as we see the intimate connection between maps and politics, maps and ethics, maps and international affairs, maps and war and peace, maps and public health, maps and climate change, maps and our yearning for a better world, maps and the attitudes we wake up with every morning... clearly seeing such connections can help bring us closer to the world we want.

P.S.

(Parting Shot)

If you picked up this book expecting it to be all about maps, you were in for a surprise, right? It moved further into politics and faith and the values we hold, into human relations, justice and peace and budgets and environmental concerns – in short, how we deal with the world – than you anticipated.

Actually, the assertion that maps have agendas, though long and vigorously resisted, is now widely accepted. But we haven't reached the end of the journey; the next phase will come as people recognize the function of values in the whole map mix. We pointed to the important role values played – and play – in the popularity of the Peters map (and, by implication, in any equal-area rendering), but values are implicit every time a map is devised. If you seek further evidence of that, reread those sections dealing with the Grand River/Caledonia controversy, the California logging map, the Energy Task Force map of Iraq, whether to set the edge of a world map in Russia, the choice of "up," among other examples, and then ask yourself what values were in play. To restate the thesis: there is no way to deal with maps without also opening up a world of other questions.

But that's the striking thing about maps. They have the power to take us into unexpected adventures. Enjoy the trip!

Endnotes

Chapter 1

1. The phrase is borrowed from *The Power of Maps* by Denis Wood. New York, The Guilford Press, 1992.

2. Groups using this book for study and discussion might consider presenting the following scenario:

   In foreign Country X a committee has been charged with recommending changes in their nation's foreign policy. Spread out before them is a map of the U.S.A. with highly selective data. Actually, it is a map tailor-made for this purpose. To illustrate how it picks and chooses from available data, we zero in on the state of Georgia. And what do we find?

   • any reference to the original inhabitants, early exploration, or Civil War sites? No!

   • data about population distribution? No! (While Atlanta is shown, Savannah, Augusta, and Athens are not mentioned.)

   • highways or airports marked? No!

   • natural attractions such as Stone Mountain? No!

   • centers of learning and public interest: Emory University, the Heard Museum, the Carter Center, the Centers for Disease Control? No way!

   • major employers such as Coca-Cola's world headquarters and the Ford Assembly plant? No again!

   Instead, the map before the task team fixates on peanuts:

   • it shows present growing areas, with figures on capacity,

   • it identifies areas that could be converted to peanut cultivation, and

   • it locates shipping ports to use to get product out of the United States and into country X, which has an amazing appetite for peanuts.

   Let your group assume a similarly selective agenda in the foreign task team's mapping of all 50 states – concentrating on oil in Texas, oranges in Florida, cheese in Wisconsin. (Canadians, Australians and others will adapt the idea to their own situation.) If the foreign team had no bias to start with, what effect might their map have on their perspective? What reaction would you expect from the target nation if and when they learned about the other nation's covert intentions? Would that be an appropriate response? To what extent does this fictional narrative shed light on the real-life story of the Task Force on Energy? What would you add to or subtract from this imaginary scenario to make it more realistic?

Chapter 2

1. *Let me get this straight, Mr. Winchester. You're saying maps are ... **propaganda?** But isn't propaganda what our enemies engage in? Aren't mapmakers supposed to be our friends?* To take seriously this statement from Simon Winchester, a respected researcher and writer, and then to examine it and question it, is to take the next step in a meaningful conversation about the way maps function in human affairs.

2. Projection is the term commonly applied to the process by which features on a spherical surface – the earth or a globe – are transferred to a plane or flat surface. A fuller explanation of the concept may be found in a number of places, including many atlases, geography textbooks, and *Seeing Through Maps*, pp. 16-19.

3. Buckminster Fuller, *Ideas and Integrities.* New York: Collier Books, 1969, p. 124.

4. Buckminster Fuller, *Fuller Projective-Transformation*, an interpretation of the Dymaxion map.

5. From the interpretation accompanying the map.

6. For ordering information, go to ODTmaps.com.

7. *Mercator's World,* March-April 1997, p. 9.

8. James Laxer, *The Border: Canada, the U.S. and Adventures Along the 49th Parallel.* Toronto: Doubleday Canada, 2003, p. 42.

9. Charles M. Johnston, ed., *The Valley of the Six Nations: A Collection of Documents on the Indian Lands of the Grand River.* Toronto: University of Toronto Press and the Champlain Society, 1964.

10. Wisconsin State Historical Society, Draper mss, 13F24, quoted in Edward S. Rogers and Donald B. Smith, Aboriginal Ontario: Historical Perspectives on the First Nations. Government of Ontario © Her Majesty the Queen, 1994, p. 170.

11. *Frommer's Vancouver and Victoria*, 2008. Hoboken, NJ: Wiley Publishing Inc., p. 8. See also "Settling Land Claims," by Mary C. Hurley (Law and Government Division) and Jill Wherrett (Political and Social Affairs Division), Library of Parliament, Ottawa, PRB 99-17E.

12. To make that statement is to raise the question whether North America's indigenous people had or created maps before the arrival of Europeans. This is currently a matter of debate among cartographers. Some have stated that maps have a long history, even predating the invention of writing. The monumental *History of Cartography*, especially Vol. 2 Book 3, Cartography in the Traditional African, American, Arctic, Australian, and Pacific Societies, by David Woodward and G. Malcom Lewis, eds. Chicago: University of Chicago Press, 1998, treats indigenous efforts as maps. Denis Wood, on the other hand, in his *Rethinking the Power of Maps*, cogently argues that maps in the modern sense did not appear until about 1500. Clearly this would set pre-Columbian efforts outside the maps category. Here we treat early indigenous efforts as useful within their own context, whether or not they contributed to the development of modern mapping.

13. A fertile field for further research – perhaps in a doctoral program – would be a comparison of the role of "talk" in indigenous spatial communications and in Western mapping. See Chapter Eight in *Seeing Through Maps* (Revised edition, Wood, Kaiser, Abramms) for a discussion of Western maps as "talk."

14. The Supreme Court ruled, in part, that oral tradition in aboriginal culture is to be valued and respected, not set aside in favor of written documents. In principle, as I see it, their judgment applies absolutely to indigenous mapping methods as we have contrasted them in this chapter with Western cartography. In explaining its ruling, the Court referred to the *Report of the Royal Commission on Aboriginal Peoples* (1966), vol. 1 (Looking Forward, Looking Back).

   "The Aboriginal tradition in the recording of history is neither linear nor steeped in the same notions of social progress and evolution [as in the non-Aboriginal tradition]. Nor is it usually human-centred in the same way as the western scientific tradition, for it does not assume that human beings are anything more than one – and not necessarily the most important – element of the natural order of the universe. Moreover, the Aboriginal tradition is an oral one, involving legends, stories and accounts handed down through the generations in oral form. It is less focused on establishing objective truth and assumes that the teller of the story is so much a part of the event being described that it would be arrogant to presume to classify or categorize the event exactly or for all time.

   "In the Aboriginal tradition the purpose of repeating oral accounts from the past is broader than the role of human history in western societies. It may be to educate the listener, to communicate aspects of culture, to socialize people...

   "Oral accounts of the past include a good deal of subjective experience. They are not simply a detached recounting ... but, rather, are 'facts enmeshed in the stories of a lifetime.' They are also likely to be rooted in particular locations, making reference to particular families and communities. This contributes to a sense that there are many [stories], each characterized in part by how a people see themselves, how they define their identity in relation to their environment, and how they express their uniqueness as a people."

The Court drove to its landmark ruling:

"Notwithstanding the challenges created by the use of oral histories as proof of historical facts, the laws of evidence must be adapted in order that this type of evidence can be accommodated and placed on an equal footing with the types of historical evidence that courts are familiar with, which largely consists of ... documents."

This brief summary, let it be noted, is part of a Court decision that runs to about 500 pages. To pursue the matter, go to Supreme Court of Canada – Decisions – Delgamuukw v. British Columbia, 1997. A provocative analysis may be found in John Ralston Saul's *A Fair Country: Telling Truths about Canada.* Toronto: Penguin Group, 2008.

15. For further information on North American indigenous mapping, good sources are *The History of Cartography*, Vol 2 Book 3, cited in footnote 2-12, *Dark Storm Moving West* by Barbara Belyea. Calgary: University of Calgary Press, 2007, and *Boundaries of Home: Mapping for Local Empowerment*, ed. By Doug Aberly. Gabriola Island, B.C. and Philadelphia, Pa.: New Society Publishers, 1993.

16. Address at Chautauqua Institution, August 20, 2004.

17. *Hegemony or Survival: America's Quest for Global Dominance*, pp. 170-171, New York: Metropolitan Books, Henry Holt, 2003. The term Bantustan, sometimes rendered "homeland," designates territory set aside for black people in South Africa and South West Africa, now Namibia. Official dogma held that the Bantustans benefited everyone; in reality they segregated blacks not only from privileged whites but from other black communities, disempowering them. In effect they followed a "divide and control" strategy.

18. In a significant sequel to the story, former president Jimmy Carter tackled this issue head on. Sharp dissent as well as appreciation ensued. Noam Chomsky, in an interview in London published in *New Internationalist*, said admiringly, "The facts did finally break into the mainstream with the publication of Jimmy Carter's *Palestine Peace Not Apartheid*. The book elicited a torrent of abuse and desperate efforts to discredit it." Chomsky then offers his perspective: "The imperial mentality is so deeply embedded in Western culture that this travesty passes without criticism, even notice."

1. An earlier version of this chapter appeared in *The Cartographic Journal*, Cambridge, U.K., 40(1) June 2003.

2. *Synchronoptische Weltgeschichte* and *Histoire Mondiale Synchronoptique* are available from ODT. A digital version was released by Huber Kartographie in 2011. www.derdigitalepeters.de

3. Letter to the author.

4. *Seeing Through Maps*, for example.

5. *The New Cartography*. Universitätsverlag Carinthia, Klagenfurt, Austria and Friendship Press, New York, 1983.

6. "Cartography's Defining Moment: The Peters Projection Controversy, 1974-1990" in *Cartographica*. University of Toronto Press, Winter 1994, pp. 16-32.

7. Cited by Stefan Mueller, "Equal Representation of Time and Space: Arno Peters' Universal History," in *History Compass*, Vol. 8, Issue 7, pp. 718-729, published online July 2, 2010. Hoboken, NJ: and Chichester, U. K., John Wiley and Sons, Inc. DOI: 10.1111/j.1478-0542.2010.00693.x

1. Mercator lived from March 5, 1512 to December 2, 1594. He launched his map in 1569. So the appearance of this book coincides with Mercator's 500th birthday!

2. *The Canadian Geographic*, Vol. 97, No. 1.

3. Letter from Rand McNally to ODT, Inc.

4. *Mercator's World*, Sept.-Oct. 1997, p. 55.

5. Other handy terms might be – Mercator's Proxies or Synoptic Maps (from Greek syn: "together," "collectively," or "alike" and optic: having to do with the eye and vision; thus Synoptic is used to mean offering a similar view.) Whatever the term, the important point is, some maps look surprisingly similar. In perpetuating to a significant degree the size distortions of the Mercator – enlarging the high latitudes while diminishing tropical areas and adopting a rectangular format – they reinforce rather than correct distorted understandings of the world.

6. Arno Peters, *The Europe-Centred Character of Our Geographic World View and Its Correction*, text of a lecture to the German Cartographical Society, Berlin, 1974. Universum Verlag, Munich-Solln, p. 9.

7. John P. Snyder and Philip M. Voxland, *An Album of Map Projections*, U. S. Geological Survey Professional Paper 1453, p. 10.

8. Simon Winchester, "The Mapmaker to Blame for Distorted Worldviews" in *The New York Times*, Jan. 22, 2002.

9. "Peters Projection vs. Mercator Projection" by Matt T. Rosenberg.

Chapter 5

1. *Seeing Through Maps: Many Ways to See the World*.

2. "Use of Cylindrical Projections for Geographical, Astronomical, and Scientific Purposes," in *The Scottish Geographical Magazine*, Edinburgh. Vol. I, 1885. Gall named his three maps the Stereographic, Isographic, and Orthographic Projections. The Orthographic is the equal-area map.

3. J. B. Harley, "Can There Be a Cartographic Ethics?" in *Cartographic Perspectives*, No. 10, Summer 1991, p. 11.

4. Letter to the author.

Chapter 6

1. For more detail, see Clark Blaise, *Time Lord: The Remarkable Canadian Who Missed His Train and Changed the World*. Toronto: Knopf Canada, 2000.

2. Mercator himself, following precedent set by Ptolemy and others, located the Prime in the Canaries.

3. Others are getting into the act. A call to relocate the Prime to Mecca came out of a conference of Muslim scientific and religious leaders, held in 2008 in Qatar. The gathering, whose conclusion seems to have been determined in advance (its title, "Mecca, the Centre of the Earth, Theory and Practice," seems a giveaway) urged the world to recognize the "centrality" of Mecca and to reset time zones accordingly. A provocative postcard-size map of the world showing Mecca as the center of the world is available from ODT, co-publishers of this book. It was created by Len Guelke, who also developed the Beamsville-centered map featured in Chapter 2. Significantly, it does not lobby to make Mecca the locus of the Prime; rather, it is a cartographically legitimate way of gaining perspective on the world.

4. Interestingly, the effective date line once ran between Alaska and what is now Canadian territory. But in 1867, when Alaska passed from Russian to American control, that section of the line was moved westward, enabling Alaskans to synchronize their calendar with the rest of the continent. And as recently as 1995 Kiribati shifted the line, ending the anomaly of that tiny country being split into two days.

5. To appreciate just how big a continent, consult an equal-area map such as the Peters or Hobo-Dyer, not one of the Mercator Look-alikes that minimize it.

6. Congress established the dollar and a decimal monetary system in 1786, so supplanting the traditional and more cumbersome British pattern.

7. For example, so notable a figure as Britain's astronomer royal, Sir George Airy, publicly ridiculed Fleming's proposals: "… as to the need of a Prime Meridian, no practical man ever wants such a thing." Airy had introduced standard time in Britain thirty years earlier, but opposed applying it to the wider world.

8. Britain's Royal Geographic Society has actually taken the first step in that direction. When Peters personally presented his proposal to them he was, in effect, inviting them to give up their home-team advantage: to move the Prime half way around the world. Yet, to the great surprise of many skeptics the RGS did not reject the idea. Their cautious but open reply: the plan was "not impossible."

Chapter 7

1. A video of the event is available on YouTube – Miss Teen USA South Carolina 2007. Other examples of geographic ignorance/map illiteracy may be cited. One that got little media attention but was perhaps more embarrassing is related by Henry Kissinger in his *Years of Renewal*. Kissinger tells how President Nixon, meeting with the Prime Minister of Mauritius, confused that country with Mauritania, based on information staff had supplied. On maps, one country is a small island in the Indian Ocean; the other, much larger, is in Africa's far west. Culturally, economically, and in their relations with the United States they were a world apart, so the conversation between two heads of state, Kissinger reports, led to frustration rather than progress.

2. Of related interest is the conflict between South Korea and Japan over a tiny group of bleak islands lying between the two countries. Though satellite imaging has not (so far) played a significant role, both countries assert rights of ownership. Competing claims have escalated to threats of violence.

3. To be more precise, there is also a third, known as the celestial Pole, determined by Polaris or the North Star, but it is of little relevance here.

4. Glenn T. Trewartha, Arthur H. Robinson and Edwin H. Hammond, *Physical Elements of Geography* (Fifth edition, 1967) New York: McGraw-Hill, p. 24.

5. *A New View of the World: A Handbook to the World Map: Peters Projection*. New York: Friendship Press, 1987; Amherst, Mass: ODT, Inc., 1993.

6. Mapmakers regularly acknowledge distortions in their creations. Thus the National Geographic Society has pointed out that the Winkel Tripel, its current world map, and its predecessor, the Robinson, distort both size and shape, but to an extent they consider acceptable.

7. *Peters World Atlas: The Earth in its True Proportions* published in Germany by Ullmann and distributed in the USA by ODTmaps.com

8. Personal conversation between the author and Jackie Robinson.

9. *Seeing Through Maps: Many Ways to See the World*. ODT, Inc. 2001, 2005, 2006.

10. A fuller summary of Saarinen's research may be found in *Seeing Through Maps*, pp. 34-35. See also Richard Monastersky, "The Warped World of Mental Maps: Students Worldwide Share a Skewed Vision of the Continents," in *Science News* 142 (October 3, 1992), pp. 222-223.

11. Since this exchange took place I have seen results of a study relating smoking on screen and teenage tobacco use. It points to a clear correlation between seeing tobacco used in movies and picking up the smoking habit. The study, published in September 2007, may be accessed at *Archives of Pediatric and Adolescent Medicine* 161 (9), pp. 849-858.

Is this relevant? Definitely, in my opinion. The function of both movies and maps is to deliver impressions to the brain; over time such exposure "normalizes" certain behaviors. In the case of maps this may be the unquestioned assumption that north naturally belongs "on top." Or the idea that a country is primarily a source of oil (or oranges or illegal immigrants) rather than a people's homeland. Or the unthinking acceptance of disparities of importance, opportunity and power between the privileged and dispossessed parts of the world.

As a result of the smoking study (supported by the National Cancer Institute and the American Legacy Foundation), the Motion Picture Association of America has acted to include smoking as one factor in rating films. Is the case of maps analogous: should some maps be rated R, Not for General Viewing?

1. For the full text and audio of Bono's remarks, go to www.americanrhetoric.com/speeches/bononationalprayerbreakfast.htm

2. From *You and the Nation's Priorities*, New York: Friendship Press, 1975, p. 16.

3. "The Agency of Mapping," in D. Cosgrove, *Mapping*, London: Reaktion Books, 1999, p. 213.

1. Reinhold Niebuhr, *The Irony of American History*. New York: Scribner, 1952. Niebuhr's references to such terms as imperialism, pretensions (arrogance), and justice/injustice may seem harsh, especially to those Americans who believe their country acts out of noble motives for the benefit of others. Nevertheless, a growing body of works by analysts attests to the contemporary relevance of Niebuhr's position. See, for example, *God and Empire: Jesus Then and Now*, by theologian/political analyst John Dominic Crossan (New York: HarperCollins, 2007), *Perils of Empire* by political scientist Monte L. Pearson (New York: Algora Publishers, 2008), *The Perils of Empire* (that's correct, same title) by another political scientist, James Laxer (Toronto: Viking Canada, 2008), *In Praise of Barbarians: Essays against Empire* by Mike Davis (Chicago: Haymarket Books, 2007), and *The Forging of the American Empire: From the Revolution to Vietnam: A History of American Imperialism* by Sidney Lens and Howard Zinn, (London: Pluto Press, 2003). In his 2010 book *Washington Rules: America's Path to Permanent War* (New York: Henry Holt) Andrew Bacevich states his conviction: American claims to be operating out of "benign purposes" are false. Bacevich writes with the perspective of a West Point graduate with 23 years of army service. (At the same time, we recognize that others strongly hold to a different view; thus Max Boot argues in *USA Today* (May 5, 2003) that "U.S. imperialism has been the greatest force for good in the world during the past century." Even so, we note that commentators on both sides of the debate clearly acknowledge that imperialism is alive and well.)

   By *imperialism* in this context I mean the modern variant of what was once called colonialism: the domination, through overpowering strength, of others by one nation, one economic or cultural system, or one faith/philosophical outlook... by *arrogance* I mean the working assumption that one's own group is so far superior that the opinions of others may be disregarded. Francis Fukuyama, political scientist at Johns Hopkins University, weighs in on that issue: what he calls "American exceptionalism" is described as the "implicit judgment that the United States is different from other countries and can be trusted to use its military power justly and wisely in ways that other powers could not." (*America at the Crossroads: Democracy, Power, and the Neoconservative Legacy*, New Haven: Yale University Press, 2006, p. 101) by *justice* I mean fair and equitable access to the world's available resources.

2. PEN is an independent, nonprofit association of writers and others concerned with freedom of expression. It assists those it deems to have been unjustly silenced under authoritarian regimes.

3. *Being Muslim*, Toronto: Groundwood Books, 2008, p. 27.

4. Many other perceptive commentators have called attention to the split or chasm in the world. The remarkable thing is the surprising convergence of their ideas, though they come

from widely varied backgrounds. Among them:

Margaret Atwood. In *The Handmaid's Tale*, this renowned novelist depicts a society under stress: if things aren't going well it must be the communists, or the Jews, or the liberals, or evil people who can't stand it that others are free … so the world gets divided into "us" – the white hats – and "them" – the evil-doers who need to be set right or stopped dead.

Martin Amis. This British novelist tells of people demeaned, stripped of dignity and power, who actually find an ultimate pleasure in killing.

John Updike in *The Terrorist*. "I felt I could understand the animosity and hatred which an Islamic believer would have for our system," explains the prize-winning author, commenting on his work.

Condeleeza Rice, United States Secretary of State in the second George W. Bush administration, who speaks of "freedom's divide" and those "unlucky enough to be born on the wrong side of that divide."

Robert Rubin. In his book *In an Uncertain World* this former Secretary of the Treasury (Clinton administration) sets out a declaration of interdependence that, if acted upon, would help bind together the peoples of the world across the present chasm of separation.

Black Gold. Using coffee as a case study, this British documentary film sharply contrasts capitalist greed vs. Third World need. It asks the basic question why, when people in the developed world pay $3 to $4 a cup at specialized coffee shops, only one or two cents is paid to those who grow the beans.

Michael Ignatieff. In *Blood and Belonging* this Harvard-academic-turned-political-leader-in-Canada-turned academic points to the *psychological* reality that underlies the chasm: "Between the hungry and the sated nations, there is an impassable barrier of incomprehension."

Richard Gwyn, London-based syndicated columnist. "To look closely at rich-poor, or North-South, trade is to look directly at a nightmare. The U.S. collects more in tariffs from its trade with Bangladesh than from its trade with France. Japan imposes 1,000 per cent duties on imported rice. Mozambique loses as much because of European sugar subsidies as its gets from European foreign aid."

The International Red Cross. Its *World Disasters Report* charges that humanitarian aid from developed nations to areas of need reflects political priorities more than an ethic of help based on human need.

Worldwatch, an independent research organization based in Washington, adds that the root cause of hunger is not shortage of food but poverty. Of all malnourished children in the Third World, 80 percent live in countries that have enough food, but it is not available to the poor.

The International Summit on Democracy, Terrorism and Security brought together more than 200 global leaders and scholars in Madrid in 2005. High on their list of conclusions was the pressing need for global justice.

Finally, **John McCain,** in a pronouncement many will find surprising – given other stances he has taken and the bellicose image he projects – stated that, in any effort to shift the relationship between the West and the Islamic world, "scholarships will be far more important than smart bombs."

5. A math teacher in a Toronto public school gave her 7th Grade students the following homework assignment:

(a) Compare the wages earned by a Chinese factory worker ($0.28 an hour) with the salary of Nike's CEO (approximately $9600 an hour)

(b) Calculate how many hours one factory worker would have to work to earn the equivalent of what Nike spends annually on advertising (950,000 years).

Would you encourage curriculum planners and teachers in your local schools to include problems of this nature? Why or why not? What new awareness might develop? What possible action?

1. Letter to Friendship Press.

2. If the church and the larger society were to take this option seriously, it would radically alter both local and global economics. In seeming recognition of that, a candidate for the Republican Party's nomination for the 2012 U.S. presidency was asked to give his opinion on this "preferential option." (He skirted the question.) To learn more, a good resource is Charles E. Curran, *Catholic Social Teaching*: 1891- Present. Washington, DC, Georgetown University Press, 2002.

3. Letter to the author.

4. Letter to the author.

5. Conversation with the author.

6. Letter to the author.

7. Letter to the author.

8. I base this statement on experience which, while extensive, is also limited. Therefore if any reader will identify any group that regularly uses equal-area maps and also maintains an intolerant or exclusivist stance, I'll revise my position at the next opportunity. And take that person to dinner!

9. "A Map for Getting Lost," *Mercator's World* Vol. 2 No. 2, March/April 1999, p. 34.

Chapter 11

1. *Western Political Thought in the Twentieth Century.* London: 1985, p. 163.

2. James Wolfensohn, speech in Shanghai, May 25, 2004.

3. Greg Mortenenson, *Stones into Schools,* New York: Viking, 2009, p. 70.
   As of this writing, Greg Mortenson is the subject of controversy; indeed, has been publicly accused of manipulating facts and mishandling contributions to the charity he has set up. Without taking a position on the merits, I suggest the questions raised point to the reality that we are all flawed. If this book were to refer exclusively to perfect persons – or to be written by a perfect author – how much of it do you suppose would exist ? Still, criticism may not damage the importance of Mortenson's insight. Even imperfect people can sometimes get things right! The same stance applies to institutions: medicine, banking, sports, journalism, religion, education, whatever. For all their shortcomings, we continue to deal with them, and rightly so.

4. The history of Germany between the two world wars may be seen as a cogent example of global failure to adopt that perspective of enlightened self-interest. The victorious Allies determined to punish the Germans for wrongs done. They exacted reparations until Germany was effectively bankrupt and popular resentment helped pave the way for Hitler and World War II. In retrospect, would the Allies have been better served by a less punitive, more humane policy? How do we draw the line between "foolish" and "enlightened" self-interest?

5. Gunther Krause and Matthias Tomczak. "Do Marine Scientists Have a Scientific View of the Earth?" in *Oceanography*, Vol. 8 No. 1, 1995, pp. 12, 13, 15. More resources on the web include: Why are map projections essential tools of physical oceanography? which requires you to go through a quiz... about 8 pages later you get to a very interesting comparison.

6. Reported in Ben F. Barton and Marthalee S. Barton, "Ideology and the Map: Toward a Postmodern Visual Design Practice," in *Central Works* in *Technical Communication*, New York: Oxford University Press, 2004, p. 239. For additional analysis of the issue, see Toxic Wastes and Race in the United States: A National Report on the Racial and Socio-Economic Characteristics of Communities Surrounding Hazardous Waste Sites. New York: United Church of Christ Commission for Racial Justice, 1987, and Mark Monmonier, Cartographies of Danger. Chicago: University of Chicago Press, 1997.

7. David Korten, *The Great Turning: From Empire to Earth Community*, 2006, San Francisco: Berrett-Koehler Publications, Inc. Used by permission.

8. *Caring for the Future: Report of the Independent Commission on Population and Quality of Life.* New York: Oxford University Press, 1996, p. 257.

Chapter 12

1. Based on Tom Koch's work; see *Cartographica* Vol. 39, #4, Winter 2004, and *Cartographies of Disease: Maps, Mapping, and Medicine.* Redlands, CA: ESRI Press, 2005.

2. It should be noted that there are professionals who object strenuously to some of Koch's conclusions. Some have claimed that Koch misconstrued Snow's map. The full discussion can be found in *Cartographica*, Vol. 40, Issue 3 (Fall, 2005). The very fact that there is debate aptly illustrates the subtitle of this book: persons engaging in a *conversation* – in his case at a professional level – about how maps make a difference. My purpose is to illustrate how mapping can serve public health. Significantly, both sides of the debate, despite their differences, point to John Snow's map and all that it helped bring about as central to the development of modern epidemiology. That is our point. For anyone seeking another perspective on the issue, a good source is http://www.ph.ucla.edu/epi/snow.html.

3. "Crime Sleuths Crack Down on Home Break-Ins Using Predictive Maps," *Directions Magazine*, May 10, 2004.

4. For additional background go to www.justicemapping.org; see also TIME, March 15, 2007, *The New York Times*, Nov. 23, 2007, and *The Toronto Star*, July 19-26, 2008.

5. Thomas Buell, Jr., Vice President, Global Pittsburgh, in letter to the author.

6. Sustainability Institute, currently known as the Donella Meadows Institute.

7. Roméo Dallaire with Brent Beardsley, *Shake Hands with the Devil: The Failure of Humanity in Rwanda*. Toronto: Random House Canada, 2003. p. 521.

8. Ibid. p.522.

9. From the BBC film *Shooting Dogs*, 2005. Released in the USA as *Beyond the Gates*.

10. Stephen Lewis, *Race against Time: Searching for Hope in AIDS-Ravaged Africa*. The Massey Lectures, November, 2005. Toronto: House of Anansi Press.

11. Speech to Amnesty USA's Annual General Meeting, Brooklyn, NY. April 16, 2004. Used by permission.

Study Guide to
How Maps Change Things

by James Taylor

Introduction

Learning Goal

a) To help participants recognize how the maps they use influence their perceptions of the world they live in, and

b) To encourage them to join in active advocacy for a more equitable world.

It's also worth identifying a few things that this study program is NOT intended to do:

- It is not a promotion for a particular kind of map.
- It is not a process for teaching a particular body of knowledge to the participants.
- It is not a means for any one person or group to demonstrate their superior knowledge of geography, culture, or economics.
- It does not propound any particular economic, political, or religious ideology.

Learning Process

Participants will combine research with personal experience through discussion in small groups. Participants will both discover and consolidate insights through work with maps.

Some may feel that coloring maps is a task better suited to children than adults. In reality, information absorbed through a single sense – through the ears in a lecture or through the eyes as a video or slide show – is only absorbed superficially. When people also use their muscles, for actions such as marking up a map, they will remember far better what they have done.

So each session will include times for participants to share their previous experiences with maps (or the lack of them, in some situations!), to research information that they may not have thought about before (or may not even have known existed), and to identify what they have learned.

Each session suggests a variety of exercises. You do not have to do them all. Look over the possibilities in advance, and decide which exercises are most likely to work well for the size and composition of your study group, and that will fit into the time available.

If time is short, do the exercises (or portions of exercises) most likely to have a lasting impact on the participants' awareness. The exercises should generate lively discussion – as long as the discussion is relevant to the theme, it's probably better to let it run than to cram in an additional exercise.

As a personal option, you may wish to provide coffee/tea/juice/water (and perhaps cookies) for each session.

The Program

There will be four sessions of about 75 minutes each – although eager groups might wish to expand some sessions as they continue to explore ideas and information. Remember, however, that it's better to leave people wanting more, and continuing discussions among themselves after the formal meeting time, rather than wearying of the subject.

Session One provides an introduction to the conventions of map-making. It will help participants identify the distortions that inevitably occur when mapmakers try to translate a spherical surface onto a flat sheet of paper. All maps distort reality; they simply choose different factors to distort. Maps are never neutral!

Session Two considers how maps affect the user's point of view by these selective distortions of size, shape, orientation, and content. Some maps focus on human factors: streets, cities, national boundaries. Others focus on natural factors: topography, resources, etc.

Session Three invites examination of some demographics that maps conventionally don't show, such as poverty, energy use, water consumption, life expectancy, religion...

Session Four will try to identify the new ideas about this world we share that may have developed during the previous three sessions, and will encourage participants to dream about how they could get involved in creating a better world.

The Facilitator's Role

Every session will require one person to organize the materials for that session and to keep the process moving. For simplicity, we refer to this person as the facilitator.

The facilitator may be male or female, young or old. The facilitator does not need to have any special knowledge of maps or geography, and should not attempt to be an authority who provides answers. The facilitator simply keeps the exploration process moving smoothly.

Many different kinds of groups may undertake this study-and-action process. Some may already have skills and expertise to contribute. Participants may also share professional, environmental, economic, or religious interests. The only special skill required of a facilitator is to discern the particular interests of the group, and to encourage discussion relevant to those interests. For example, business people might wish to focus on how maps reflect economic issues; a church group might consider how maps relate to their religious convictions.

Rules of Order

Because no decisions are being made in these sessions, there is no need for formal procedures such as Roberts' Rules of Order (in Canada, the authority is Bourinot, but the principles in both systems are similar). However, it is worth pointing out some general guidelines for courteous and open discussion. You may adapt or amend these suggestions; you might outline them orally or print them for distribution.

- Every participant is entitled to an opinion.
- Opinions based on personal experience and observation are preferred to those that merely cite an external authority (such as, for example, a teacher, a textbook, or the Bible).
- Opinions derived from experience and observation must be considered valid for that person; opinions may be questioned, but not contradicted.
- Personal abuse or disparagement will not be tolerated.
- Consensus is not required; members may disagree on what they have learned from the various sessions and exercises.

What You Will Need

- A globe of the world, preferably one showing topography (land forms) rather than nation/state boundaries. If you don't own, or can't borrow, a globe, buy one – it's worth it! You can order a globe through EarthBalls (http://www.earthball.com/orders.htm) or from ODTmaps (http://www.odtmaps.com/detail.asp_Q_product_id_E_earthball-16inch). Transparent inflatable globes are also available through the ODTmaps page.
- A good atlas, with an index for locating little known countries and places.
- An Internet connection (wireless or hardwired) so that participants can use laptops and/or tablets to do searches, use apps, and access websites.
- At least one participant should have a laptop computer or tablet – with a screen big enough to be seen by more than one person at a time.

- Some coloring tools (crayons, markers, pencils) in at least four colors: red, yellow, green, and blue.
- A selection of handouts (included with each session) photocopied for use by participants.
- One ball, without markings that might imply a top or a bottom end.
- A flip chart or equivalent (blackboard, whiteboard, etc.) for writing brainstormed suggestions on (for Session Three only).
- A stopwatch or kitchen timer (for Session Four only).

Study Guide to *How Maps Change Things*

Learning Goal:
To discover how maps can distort reality.

Materials needed:

- Copies of the book *How Maps Change Things*. Every participant should have one.
- A questionnaire (Handout #1, at the end of this session) with copies for everyone.
- A map of the world, Mercator projection (Handout #2) with copies for everyone.
- A globe of the world.
- A piece of string long enough to go at least halfway around the globe. (A small, very flexible tape measure would also work, but avoid straight rulers and the kind of heavy-duty retractable measuring tapes favored by carpenters.)
- A laptop computer and/or tablet connected to an Internet server.

Ensure that participants can sit at a table. If the number of participants is greater than about eight, consider dividing the group into two, each with a full set of tools and materials. The ideal group will be no less than four and no more than eight persons.

Getting started

Distribute copies of the book *How Maps Change Things* by Ward Kaiser.

Ask participants to introduce themselves, even if they already know each other. Invite each person to explain briefly why they are interested in taking this study program about maps.

(estimated time, up to 10 minutes)

Considering the purpose of maps

Hand out copies of the questionnaire (Handout # 1, provided at the end of this session). Have each participant mark the questionnaire individually, then compare their answers in twos or threes.

Ask each small group to summarize the matters on which they agreed or disagreed. Ask follow-up questions for explanations if appropriate.

(estimated time: 12 minutes)

Draw attention to the illustrations on page 28 in *How Maps Change Things*, and summarize briefly the main points:

1. Maps are flat; the world is round.
2. Any attempt to translate a round surface to a flat surface results in distortions.
3. Historically, the most common "projection" onto a flat surface has been the Mercator projection, invented by Gerhard Kremer in the 1500s.
4. Though it seriously distorts the relative size of land areas overall, the Mercator projection has the advantage of showing local land shapes accurately. (Kremer's goal was to create a map that would allow sailors to navigate the open seas with relative safety. By using a rectangular grid (east-west lines always intersecting north-south lines at right angles) he enabled navigators to calculate compass bearings and plot their course far more accurately than was previously possible. Kremer's map also placed north at the top, helping set the dominant practice for centuries.)

(estimated time: 5 minutes)

Discovering some distortions

Distribute copies of the Mercator projection map of the world provided as Handout #2 at the end of this session plan. Have participants find and mark the location of Rio de Janeiro, Johannesburg, Anchorage, Moscow, Vancouver (B.C., Canada), and Tokyo on the map.

In groups of not more than eight persons each, undertake as many of the following exercises as time permits.

Exercise 1

On the Mercator projection map of the world (Handout #2) use the string or small tape to measure the distance between Rio de Janeiro in Brazil and Johannesburg in South Africa. Now measure the distance between Anchorage in Alaska and Moscow in Russia. Which distance is greater? By what proportion?

Now repeat the exercise by measuring the distances between the same points on the globe. (Note that the shortest distance between Anchorage and Moscow goes over the North Pole, not around an east-west line.) Now how do the two distances compare to each other?

(Note to facilitator: Do not use Google or other search engines for computing these distances.)

If time permits, participants could also compare the distance between other cities on the Mercator map and on the globe – in both east-west and north-south directions.

For example, draw a straight line on the Mercator projection world map between Vancouver, B.C., and Tokyo in Japan. Does that straight line indicate the shortest distance between the two points?

Now use the string to find the shortest distance between the same two points on the globe. Does it follow the same route as on the flat map? Does this explain why flights between Vancouver and Tokyo typically take a circular route over the Aleutian Islands?

Use the string (or small tape) to check out some other "great circle" routes between North American and European or Asian cities. Do they bear any relationship with what people would expect, looking at a Mercator projection?

(estimated time: 10 minutes)

Exercise 2

Using two of the Mercator projection maps, overlay the island of Greenland on the continent of Africa. (If you have scissors available, participants could cut Greenland out of one map, to superimpose on Africa.) How do their sizes compare?

Meanwhile, have the person with the computer look up the total land area of Greenland and of Africa. How do they compare? (You'll find answers on page 21 of *How Maps Change Things*; Africa is 11.6 million square miles, roughly 14 times as big as Greenland, at 0.8 million square miles.)

You might ask questions such as

- Can a map that distorts land areas so dramatically be called "accurate"?
- How does this kind of distortion affect our perceptions of the importance of Canada or Russia, compared to India or South Africa?

(estimated time: 5 minutes)

Exercise 3

Finally, using tape or string, measure the distance from the equator to the North Pole on the globe. Do the same with the distance from the equator to the South Pole. Are the distances equal? (The word "equator" derives from "equal," meaning that this imaginary line around the earth is an equal distance from each pole.)

Now do the same measurements on the Mercator map. How much closer is the equator on the map to the South Pole than the North? How might this discrepancy affect our sense of our own importance in the northern hemisphere?

(estimated time: 5 minutes)

Discussion

Have the participants look again at the questionnaire that they marked up earlier. How many of the factors that they expected a map to provide would they still expect? Which ones no longer apply?

(Estimated time: 10-15 minutes)

Point out that the Mercator projection is not the only form of world map. There are other projections. Draw attention to pages 19-28 and 56-58 in *How Maps Change Things*. Each different form of map has its own strengths and its own weaknesses. These trade-offs must be balanced against the map's objectives: accurate lines of compass bearings, accurate area for size comparison, aesthetically pleasing images, etc.

- The Peters and the Hobo-Dyer projections represent the areas of land masses more accurately, in relation to each other, while distorting shapes and distances.
- The Robinson Projection (page 76) and the van der Grinten (page 76) are compromise projections between the Mercator's rectangular compass-bearing grid and the equal-area qualities of the Peters and Hobo-Dyer.
- The Fuller Projection converts into a reasonable model of a spherical world, but looks confusing in flat form.
- Goode's Homolosine (page 69) could also be assembled in a spherical format. It shows a truer image of the shapes and sizes of the continents, by splitting the oceans apart.
- The CIA World Factbook map <https://www.cia.gov/library/publications/the-world-factbook/maps/maptemplate_xx.html> shows the world (in an equal-area format) from both north and south at once in a very interesting manner!

(estimated time: 5 minutes)

Which of these projections do participants find most appealing? Most informative?

Invite each participant to identify one thing that they have discovered about maps that they had not previously realized.

(estimated time: 5 minutes)

Closing

Assign reading in preparation for the next session, Chapters 1-4 in *How Maps Change Things*. Announce the date and time of the next session.

If you provided refreshments for this first session, ask for a volunteer to provide similar refreshments for the next session.

Questionnaire

Which of the following items do you expect a map to indicate accurately (listed alphabetically to avoid any implied priorities)

☐   Demographics (population densities, income/poverty patterns, health)

☐   Directions (north/south, east/west)

☐   Infrastructure (roads, railways, canals, etc.)

☐   Land area

☐   Land forms (mountains, rivers, forests, etc.)

☐   Major landmarks or tourist attractions

☐   Place names (cities, streets, countries)

☐   Political boundaries (province/state, country)

☐   Proportionate distances between points

☐   Resources (geology, mining, etc.)

☐   River systems and watersheds

☐   Urban areas

☐   Other _____ (please specify)

Handout #2

Page for photocopying and distribution

Prepared for use in Jim Taylor's study guide for Ward Kaiser's How Maps Change Things

WORLD'S LARGEST CITIES*

★ Plus Anchorage

Sixty-six largest cities by population, plus Anchorage
See the full list at http://en.wikipedia.org/wiki/list of cities proper by population
On Mercator projection

Anchorage

Los Angeles
Mexico City
New York
Lima
Bogota
Santiago
Rio de Janeiro
Sao Paulo

London
Madrid
Casablanca
Berlin
Saint Petersburg
Moscow
Istanbul
Ankara
Baghdad
Cairo
Alexandria
Riyadh
Jeddah
Tehran
Addis Ababa
Lagos
Abidjan
Kinshasa
Nairobi
Johannesburg
Cape Town
Durban

Lahore
Delhi
Jaipur
Karachi
Ahmadabad
Surat
Mumbai
Pune
Hyderabad
Bangalore
Chennai

Shenyang
Pyongyang
Seoul
Busan
Beijing
Tianjin
Suzhou
Chongqing
Dhaka
Kolkata
Tokyo
Yokohama
Shanghai
Wuhan
Taipei
Dongguan
Guangzhou
Shenzhen
Hong Kong
Hanoi
Yangon
Manila
Ho Chi Minh City
Bangkok
Singapore
Jakarta

Learning Goal:

To identify some of the ways in which different kinds of maps shape our perceptions of the world we live in.

Tools and materials needed

- The globe
- The laptop/tablet with an Internet connection
- Copies of the book *How Maps Change Things.* Every participant should have one.
- The atlas
- Photocopies of the necessary maps (Handout #3, at end of session plan)
- Any large ball, such as a soccer ball, beach ball, even a tennis ball (but preferably not a basketball, which often has inscribed lines that suggest a top and bottom).

Opening

If there are any new members, go through another round of introductions. Ask those who were at Session One to bring newcomers up to date by summarizing their learnings from the previous session.

(estimated time: 10 minutes)

Homework

Invite any responses to the chapters the participants were asked to read for this session. Were there any significant discoveries, pro or con? Any surprises? Did any particular comments or illustrations leap out at them?

(estimated time: 5 minutes)

Introductory explanation

Remind participants that all of the maps they looked at last week (even the Fuller projection) tend to show north at the top. Show the ball.

- Where is the top of a ball? Invite discussion.
- Why do we almost universally make north the top of our globes and maps?
- How does this choice influence our attitudes towards the rest of the world?

(estimated time: 5 minutes)

Exercise 1

Draw attention to the two maps on pages 27-28 in *How Maps Change Things.* Note especially what happens when south goes on top. Or when the focus is on oceans rather than land masses.

In advance, cut one of the maps and reassemble it, with glue or tape, so that the Americas are centered, and Asia/Russia are split in half. Pass it around. How does this change our perceptions of North American relations with the rest of the world?

If the globe is moveable, invite each participant to hold the globe so that his/her eyes are directly above Tahiti in the South Pacific, an ocean which occupies almost half of the earth's surface. Without moving the globe, how much of the world's landforms (continents) can they actually see?

Try also looking at the globe from directly over the North Pole. Or from directly over the South Pole. How different does the world look? What disappears from your awareness?

Invite discussion about why we choose to organize maps the way we do.

(estimated time: 15 minutes)

Exercise 2

Have each person identify his/her birthplace on the globe(s). (If you have more than one working group, you will need a globe for each group.) In each case, rotate the globe (if possible) to put that person's birthplace at the top, or at least directly in line with the owner's eyes.

How does making one's birthplace the center of the world affect one's perceptions of the rest of the world? How do the other parts of the world rearrange themselves according to this perspective? Discuss their discoveries in small groups.

(estimated time: 10 minutes)

Optional Discussion:

Ask those who have GPS devices, or navigation systems in their cars, how they orient those displays. Do they find it easier to have north always pointing up? Or is it preferable to set "up" as the direction they are going? Why?

(estimated time: 7 minutes)

Exercise 3

Color some maps according to their affiliations – that is, their colonial, linguistic, religious, or political ties. (If that seems confusing, actually doing the maps will probably make the connections clear.)

How many of these mapping tasks you can undertake will depend on the number of participants and their energy levels. If you have just a single group working together, you may only be able to assign two or three of the suggested tasks; if you have several groups, you may be able to distribute a greater selection of tasks among them. Please note – there is no need to do all of these tasks. Even a few will convey a sense of how maps shape our understandings.

You will probably need at least two maps per person, maybe more.

On a map sheet (Handout #3, Peters projection, at the end of this lesson plan) use crayons, pencils, or markers (or, for those who are especially computer savvy, Photoshop or some equivalent), to color

a) the nations/countries/territories of the former British empire. (For an example, see page 43 in *How Maps Change Things*. Use the atlas index to find less well-known countries.)

b) the countries currently hosting a U.S. military base

c) the countries currently considered to be democracies

d) the countries considered predominantly Christian

e) the countries considered predominantly Islamic

f) the countries of the G7 and the G20

g) the countries where English is the official national language

h) the countries where Spanish or Portuguese is the primary language

Handout #4 (provided at the end of this session plan) includes some suggested web links for researching lists of these factors.

(estimated time: 25 minutes)

Invite discussion of the following questions:

- As you compare these maps, what correlations do you observe?
- Which linguistic and economic factors seem to overlap? Which ones differ?
- How does looking at these correlations shape your perception of who's on top in today's world?

(estimated time: 5 minutes)

Keep these maps for future reference – they may have further correlations with coming exercises.

Closing

Assign Chapters 6-8 in *How Maps Change Things* as reading for the next session. Indicate that the activities suggested for Session Three will involve extra Internet research, so invite people to bring additional laptops or tablets for that purpose. Announce the date and time of the next session.

Ask for a volunteer to provide refreshments for the next session.

Handout #3

Page for photocopying and distribution

a) the nations/countries/territories of the former British empire. (For an example, see page 43 in *How Maps Change Things*. A list of the countries of the former British Empire and the current Commonwealth of Nations can be found at http://en.wikipedia.org/wiki/Member_states_of_the_Commonwealth_of_Nations#Current_members. Use the atlas index to find less well known countries.)

b) the countries currently hosting a U.S. military base (http://en.wikipedia.org/wiki/List_of_United_States_military_bases)

c) the countries currently considered to be democracies (http://answers.yahoo.com/question/index?qid=20060823144040AA67uJC or http://en.wikipedia.org/wiki/File:Democracy_Index_2011_green_and_red.svg or http://www.economist.com/node/8908438 or Google countries considered democracies)

d) the countries considered predominantly Christian (http://en.wikipedia.org/wiki/Christianity_by_country or Google Christian countries)

e) the countries considered predominantly Islamic (http://en.wikipedia.org/wiki/List_of_Muslim-majority_countries or Google Islamic countries)

f) the countries of the G7 and the G20 (http://wiki.answers.com/Q/Which_countries_are_members_of_the_G20 or http://en.wikipedia.org/wiki/G-20_major_economies or Google G20 countries)

g) the countries where English is the official national language (http://en.wikipedia.org/wiki/List_of_countries_where_English_is_an_official_language)

h) the countries where Spanish or Portuguese is the primary language (http://en.wikipedia.org/wiki/List_of_countries_where_Spanish_is_an_official_language (For Portuguese, use the same URL but substitute "Portuguese" for "Spanish")

Learning Goal:

To identify other human (demographic) factors that could be shown on maps, but usually aren't.

Materials needed

- Copies of the book *How Maps Change Things*. Every participant should have one.
- Photocopied maps of the Peters projection (Handout #3, from the previous session), probably two per person.
- The atlas with index
- A number of laptop and/or tablet computers for research purposes.
- A flipchart, whiteboard, blackboard, or other device for writing on, big enough that all members can read anything written on it.

Opening:

Welcome any newcomers. Provide a brief explanation of the studies that have led to this point, but don't go through the full introduction process used in Sessions One and Two.

(estimated time: 5 minutes)

Introductory comments

Note that of all the projections we have seen, the Peters projection is the one that most overtly has a purpose – to fairly present the world in a form that equalizes the land areas of the world's nations. It also deliberately uses colors to represent ethnic and regional relationships. Draw attention to pages 39 and 69-70 in the book – the text explanation for the Peters maps on page 39 is on page 69 – and compare the colors used on the Peters map to those on the Lambert map (p. 70). Can anyone see a reason for the choices of colors on the Lambert map? Do the colors bear any relationship to a nation's language or racial connections?

You could also look at the choices of colors for the Hobo-Dyer maps on page 27 of Ward Kaiser's book. Again, what connections, if any, do the colors convey?

(estimated time: 5 minutes)

Homework

Ask for any responses to the chapters the participants were asked to read before this session. What discoveries or insights can they report? Were there any surprises? Did any particular comments or illustrations leap out at them?

(estimated time: 5 minutes)

Brainstorming

What are some factors that are important to us that we have not touched upon in our earlier discussions about maps?

Use a flipchart or whiteboard (or any equivalent that can be visible to all participants) to list suggestions. You may wish to edit the suggestions from participants so that they will fit more closely to what you're prepared for! You may also add your own suggestions.

Suggestions might include (listed alphabetically to avoid prioritizing)…

- Arms expenditures
- Crime rates (per capita murders, violent crime, rapes, etc.)
- Drought, water shortages
- Energy consumption
- Energy resources (fossil fuel reserves, see map page 14 as example)
- Happiness
- Hunger/famine
- Income disparity (also called the Gini index)
- Life expectancy
- Literacy (education)
- Per capita income (wealth and poverty)
- Water consumption

(estimated time to brainstorm factors
worth pursuing: 10 minutes)

Activity

Divide into as many groups as you have research laptops/tablets – but with at least two persons in any group. Don't let individuals work alone; the discussion that takes place while doing the assigned task is just as important as completing the task itself.

Have each working group choose one of the suggestions on the flip chart list. They can use Google or any other search process to identify the best and worst in that category. (Provide Handout #5 to assist their searches.) They should then color their map to

reveal that range – using colors to denote relationships, as Peters did, rather than differences.

Coloring these maps will require a range of colors. For the sake of uniformity, we suggest red for the most serious situations (drought, famine, illiteracy, short lifespans, etc.) scaling down through yellow and blue to green for the most prosperous conditions (ample water and food, education, health, etc.). If people have limited time, they can concentrate simply on the highest and lowest in their category of concern, and ignore the mid-scale values.

Groups that finish sooner than others – whether because of facility with search engines or simpler topics – should pick an additional suggestion from the list and color an extra map.

(estimated time: 30 minutes)

Draw attention to page 86 in *How Maps Change Things*. Note the dividing line and color change on the covers of the *North-South* book that divides the globe into rich/poor, powerful/powerless, affluent/Third World…

How do the maps you colored support or conflict with this division?

(estimated time: 10 minutes)

Open discussion:

How comfortable do you feel about what you're learning about the world through these mapping exercises?

(estimated time: 10-15 minutes)

Closing

Assign reading of Chapters 8-12 in *How Maps Change Things* in preparation for the final session. Announce the date and time of the session.

Ask for a volunteer to provide refreshments for the final session.

Handout #5

Page for photocopying and distribution

List of possible sources to research for information

- Arms expenditures http://en.wikipedia.org/wiki/List_of_countries_by_military_expenditures http://www.nationmaster.com/graph/mil_exp_dol_fig_percap-expenditures-dollar-figure-per-capita, or Google arms expenditures country

- Crime rates (per capita murders, violent crime, etc.) http://en.wikipedia.org/wiki/List_of_countries_by_intentional_homicide_rate, or Google per capita violent crime rates world

- Drought, water shortages http://wiki.answers.com/Q/Which_countries_have_drought_problems, http://www.nationmaster.com/graph/hea_wat_ava-health-water-availability

- Energy consumption http://en.wikipedia.org/wiki/List_of_countries_by_energy_consumption_per_capita, or Google per capita energy consumption

- Energy resources (fossil fuel reserves, see map page 14 as example) http://www.absoluteastronomy.com/topics/Oil_reserves or http://blackjackoak.wordpress.com/2011/04/07/which-country-has-the-most-fossil-fuel/, or Google fossil fuel reserves by country

- Happiness http://www.nationmaster.com/graph/lif_hap_net-lifestyle-happiness-net, or Google happiness rates world

- Hunger/famine http://www.guardian.co.uk/global-development/datablog/2010/oct/11/global-hunger-index or http://www.fao.org/hunger/en/, or Google world hunger

- Income disparity (also called the Gini index) http://en.wikipedia.org/wiki/List_of_countries_by_income_equality, or Google income inequality countries

- Life expectancies http://en.wikipedia.org/wiki/List_of_countries_by_life_expectancy, or Google life expectancy world

- Literacy (education) http://en.wikipedia.org/wiki/List_of_countries_by_literacy_rate, or Google literacy rates world

- Per capita income (wealth and poverty) http://en.wikipedia.org/wiki/List_of_countries_by_GDP_%28nominal%29_per_capita or Google per capita income world

- Water consumption http://chartsbin.com/view/1455, or Google water consumption

Wikipedia, NationMaster, and the CIA World Factbook are among the most reliable sources.

Study Guide to *How Maps Change Things*

Learning Goal:

To do some dreaming about what would make a better world, and to imagine how those dreams might be visualized in maps.

Materials needed

- Photocopies of the questionnaire with which you started Session One.
- A printout of the possible value statements in Exercise 2 (below) to enable a tally of responses.
- A stopwatch or kitchen timer.

Opening:

Welcome participants. Remind them that this is the last session. Assure them that you will not introduce any new maps this time around!

Explain that the purpose of this session is to consolidate what they have learned in previous sessions, and to see where we go from here.

(estimated time: 3 minutes)

Homework

Ask for any comments on the chapters the participants were asked to read, prior to this session. What significant discoveries can they report? Did any particular comments or illustrations leap out at them?

(estimated time: 5 minutes)

Exercise 1

Hand out copies of the questionnaire you circulated in Session One. Ask participants individually to mark it again, based on their experiences of the last three sessions.

In twos or threes, have them discuss how their views may have changed, and what factors they may have added or deleted.

(estimated time: 10 minutes)

Exercise 2

Explain that as participants have worked with maps over the last three sessions, they may have developed some value statements – impressions of how the world works. Or doesn't work.

As you read some statements that emerged from previous studies, invite a show of agreement by raising both hands; partial agreement by raising just one hand; and strong disagreement by raising no hands. (For the time being, please don't applaud, boo, or express any other form of disagreement or approval.)

One member could keep a rough record of the group's consensus, if any, and feed that observation back to the group.

- The world is there for those who choose to make use of it. I'm entitled to whatever benefit I can derive from it.
- There is too much inequality in the world. I want to do everything I can to reduce those inequalities.
- The world is much bigger and more complex than I had realized. I feel humbled.
- The world is divided into factions – us and them. I prefer to keep my side on top.

- All the advances in education and standards of living have come from the northern nations. Everyone else should get on board and quit complaining.
- The only way to get changes made is to impose them. Power is what matters.
- Maps are not neutral expressions of fact. In future, I will know to choose a map projection that suits my needs.

Are there any other insights that have occurred to you during this study process? (Again, members may indicate agreement or disagreement, but without confrontation.)

(estimated time: 15 minutes)

Exercise 3

Assuming that the exercise above has identified some members who hold fairly divergent viewpoints, select two people to set up a short debate, using the maps they have created over the last three sessions as their primary documentation. Each person gets three minutes to state his/her view. Then each person gets one minute to counter the arguments presented by his/her opponent, again using maps as evidence.

Audience participation (in the form of heckling or applause) is welcomed during this debate, and in questions after it.

(estimated time: 10-15 minutes)

If time permits

Read the "parable" by Rabbi David Polish on page 83. Draw attention to the statistics in "The State of the Village" report on page 118.

Closing

Go around the circle, inviting each person to name the most significant thing he/she has learned during this study process. One way of ensuring that there are no challenges or interruptions is to pass a rock (or a stick, ball, or feather) from person to person – the person with the rock has the floor, until he/she passes it on. (North American aboriginal peoples often used a "talking stick" for this purpose.)

(estimated time, up to 15 minutes)

The closing itself might simply be a time for people to shake hands and chat. If you think a ritual might be suitable, you could invite everyone to gather around the globe (or globes, if you're using more than one) and rest their hands on it while you say something like this (Feel free to amend the language to suit your group; this is not copyrighted.)

Precious earth,
Blue and white planet floating
    in the immense darkness of space.
Thank you for being our home.
We are sorry for the ways
    we have mistreated you
        and all those beings – both human
            and other-than-human –
who live on your surface
    and in the depths of your oceans.
We cannot undo what has been done to you,
    whether through good intentions
        or willful ignorance.
But we pledge ourselves
    to see you differently.
We will not intentionally distort
    our understandings of you
        to suit our own interests and prejudices.
We will attempt to correct distortions
    when we hear them uttered by others.
We will work together to create a better world
    for everyone and everything.
We are part of you, and you are part of us.
So be it.

ACKNOWLEDGMENTS

Collaboration. Synergy.

That's what this book is all about. If you visualize an author sitting in solitary, please balance that image with a more relevant reality: the book in your hands owes much to many. Heading the list is the late Arno Peters, whom I came to know in 1983 and with whom I kept up a lively friendship until his death in 2002. The challenge of his vision continues, even grows, through the impressive range of his creative work and through the people in many countries he has influenced.

Particular thanks to a talented sextet of colleagues: Bob Abramms of ODT, first a customer, then my editor, publisher, co-visionary, and at all times a friend; Terry Hardaker of Oxford Cartographers, for his technical expertise and early support; Penny Watson, who became Oxford's chief cartographer and a strong source of help when Terry turned his energies to archaeology; the late John P. Snyder, whose encyclopedic grasp of mapping never overshadowed his gentle spirit; Denis Wood, whose fresh insights and outside-the-box thinking regularly challenge ideas many of us take for granted; and Mike Schwartzentruber, President of Wood Lake Publishing, for his high standards and good judgment evidenced on a daily basis, but especially in his decision to publish the book you now hold in your hands.

Arthur Bauer first voiced the need for a map handbook while he served on the board of Friendship Press; he continues his clear commitment to the Peters map and the kind of world it teases us toward. Advance readers of this book as well as those who have commented on *A New View of the World* and *Seeing Through Maps* have helped me sharpen the focus of these pages. Special thanks to Ann Hopkins, my editor at ODT, and to Brian and Simon Loffler of New Internationalist (Australia), who made even the strange new world of eBook publishing seem friendly! Family members continue to engage me in dialogue. To Sue, Gary, Chris, and Jackie and their helpful entourage I say this is a better book because of you. My wife, Lorraine, exercises a most valued function: when I get near-totally immersed – lost, let's say – in maps and their meanings she issues a gentle recall to another world – equally real – a world of food and fun and even taking out the garbage. I need that!

Appreciation to Jim Taylor for preparing the Study Guide included with this book. When feedback from the e-book version pointed to its potential in group settings, Jim Taylor's expertise commended him. If you use this book group study and action, you'll find his contribution helpful.

Finally, it has been my great privilege to lecture and lead workshops in colleges, before professional associations such as geographers, teachers, art instructors and religious educators and at meetings of service clubs and international development organizations. The Visiting Scholars program at New York University, continuing education offerings at Princeton Theological Seminary and the New York City Federation of Teachers, invitations from the Schomburg Center for Research in Black Culture and the Amy Lectures at Church of the Master/Otterbein University merit special mention. TV and radio hosts and newspaper columnists have successfully stirred public interest. From all such opportunities I continue to learn; to all these and more I remain in debt. Along with appreciation, a word of caution: others are not responsible for whatever shortcomings you may find in what you read. Those are mine alone. That's where collaboration stops!

About the Author

How do maps function?

How can we together build a better world?

These two questions, seemingly so sharply separated, come together creatively in this book by Ward Kaiser.

To his extensive experience in working for social justice he brings years of working with maps; this mix is enriched in turn by what he has done in cross cultural concerns, teaching, writing, publishing and pastoral work.

His analysis and provocative views have resulted in teaching or being published or interviewed from Bangkok to Boston and Berlin, and in Canada, Australia and the UK. Kaiser and his wife make their home at Latitude 28° 33' 37" N. and Longitude 81° 35' 3" W. and at Latitude 43° 10' 30" N. Longitude 79° 28' 40" W. , otherwise identified as on the peninsulas of Florida and Niagara.

Photo by Richard Gardner.

Also by Ward Kaiser
- The Challenge of a Closer Moon
- Focus: The Changing City
- You and the Nation's Priorities (with Charles P. Lutz)
- Launching Pad: Literacy
- Canada: A Study-Action Manual
- Intersection: Where School and Faith Meet
- People and Systems (general editor)
- Forum: Religious Faith Speaks to American Issues (contributor)
- The New Cartography (co-translator)
- Time and Space (co-translator)
- A New View of the World
- Seeing Through Maps: Many Ways to See the World (with Denis Wood and Bob Abramms)

About New Internationalist

The *New Internationalist* is an independent monthly not-for-profit magazine that reports on action for global justice. We believe in putting people before profit, in climate justice, tax justice, equality, social responsibility and human rights for all. We support minority groups and producers by selling their Fair Trade and organic products in our shop, and also like to blog and keep you in the loop via monthly email newsletters.

The *New Internationalist* also publishes a range of books on the hidden bias that favours the wealthy and powerful, including the popular No-Nonsense Guide series, available in both print and as eBooks. eBooks can be purchased from the New Internationalist online stores in Australia, North America and the UK.

How Maps Change Things is the first title in the MapAware series to be published by *New Internationalist* and *ODTmaps.com*. MapAware continues a long tradition of exposing the hidden bias inherent in traditional world maps. In 1985 the *New Internationalist* magazine published an English-language translation of Arno Peters' remarkable redrawing of the world map, and since then tens of thousands of *New Internationalist* subscribers have received a copy of the map as a free gift with their new subscription; a great introduction to "seeing the world through fresh eyes." The "Map Wars" story was revisited in the 1989 special issue of the *New Internationalist* magazine and subsequently New Internationalist and *ODTmaps.com* have cooperated to produce a range of

DVDs and books on the subject. The MapAware series builds on that history to create a contemporary library of resources that help us understand the forces that shape our perceptions and worldview.

To find out more about the *New Internationalist*, visit our Australian website at www.newint.com.au or the International magazine website at www.newint.org.

About ODTmaps.com

ODTmaps began as a management consulting company focusing on topics of employee empowerment, performance appraisal, and self-directed work teams. In 1989, ODT pioneered the use of the Peters Projection map materials in a variety of corporate culture change, leadership, and diversity diagnostic projects. When the award-winning TV show *West Wing* contacted ODT in February 2001, for permission to feature the Peters Projection map on an episode, ODT was catapulted into map publishing. Since that time, the organization has published the What's Up? South! world map, the Hobo-Dyer world map, the Population map, the incisive and insightful book, *Seeing Through Maps* and two fascinating explorations of maps and cartographers on DVD - *Many Ways to See the World* and *Arno Peters: Radical Map, Remarkable Man*.

ODT is a small employee-owned publishing company producing innovative maps and media literacy resources expanding people's worldviews. They also create custom maps for a variety of institutional and corporate clients. ODTmaps.com is based in Amherst, Masachusetts, USA and can be reached at 413-549-1293 or odtstore@odt.org

If you liked this book, post your comments on our Facebook pages:

About Wood Lake Publishing Inc.

woodlakebooks.com

Wood Lake Publishing Inc. – under its Copper-House, Northstone, and WoodLake imprints – has a 30-year history of bringing readers and faith formation practitioners unique and accessible resources that nurture, inspire, and challenge.

Our mission, undertaken through publishing, is to retrieve, reclaim, and renew the Christian tradition of living radical and inclusive love.

We are passionate about supporting and encouraging an emerging form of Christianity that is rooted in ancient wisdom and attentive to the movement of spirit in our day. This way is grounded in tradition and requires us to be dedicated to spiritual practice, and to living out our values in the world.

We are open and inclusive, and honour the perennial wisdom found in scripture and in all of the world's enduring religions.

We affirm the equality of the sexes and the godliness of people of all ages, races, and nationalities.

We understand that humans do not stand apart from creation, but exist within a web of being that is sacred in all its aspects.

We emphasize a spirituality of transformation rather than adherence to doctrine or belief.

We are pleased and honoured, therefore, to be the North American publishers of the print edition of *How Maps Change Things*.

Supplemental Full-Page Maps

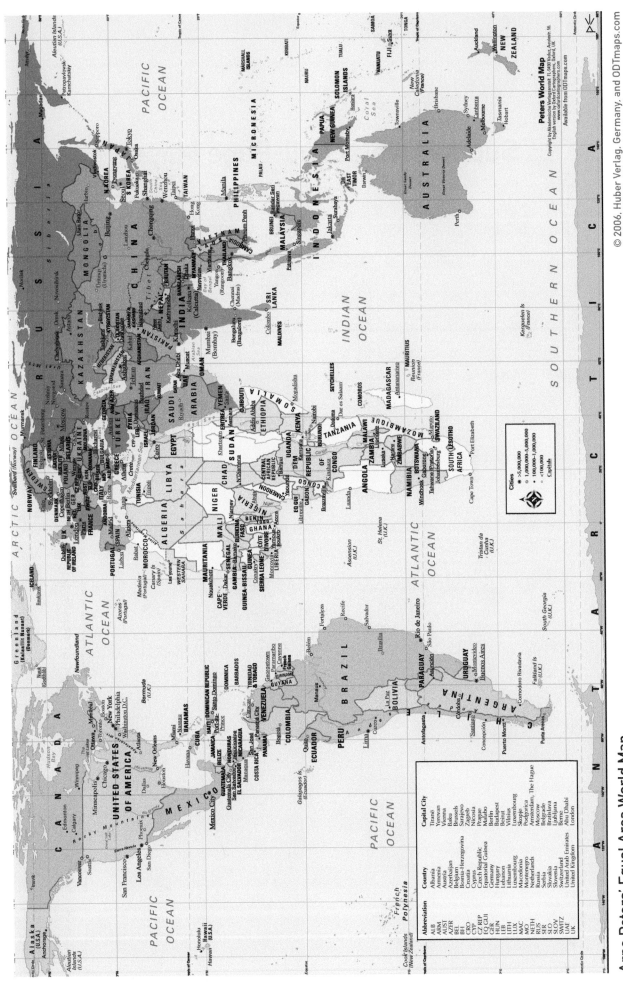

Arno Peters' Equal Area World Map

Peters World Map

Copyright by Akademische Verlagsanstalt FL-9490 Vaduz, Aeulestr. 56,
www.oxfordcartographers.com
English version by Oxford Cartographers, Oxford, UK
www.oxfordcartographers.com

Available from ODTmaps.com

PACIFIC OCEAN

ATLANTIC OCEAN

INDIAN OCEAN

SOUTHERN OCEAN

ANTARCTICA

Cities
>5,000,000
1,000,000–5,000,000
100,000–1,000,000
<100,000
Capitals

Abbreviation	Country	Capital City
ALB	Albania	Tiranë
ARM	Armenia	Yerevan
AUST	Austria	Vienna
AZER	Azerbaijan	Baku
BEL	Belgium	Brussels
BH	Bosnia-Herzegovina	Sarajevo
CRO	Croatia	Zagreb
CYP	Cyprus	Nicosia
CZ REP	Czech Republic	Prague
EQ GUI	Equatorial Guinea	Malabo
GER	Germany	Berlin
HUN	Hungary	Budapest
LEB	Lebanon	Beirut
LITH	Lithuania	Vilnius
LUX	Luxembourg	Luxembourg
MAC	Macedonia	Skopje
MO	Montenegro	Podgorica
NETH	Netherlands	Amsterdam, The Hague
RUS	Russia	Moscow
SER	Serbia	Belgrade
SLO	Slovakia	Bratislava
SLOV	Slovenia	Ljubljana
SWITZ	Switzerland	Berne
UAE	United Arab Emirates	Abu Dhabi
UK	United Kingdom	London

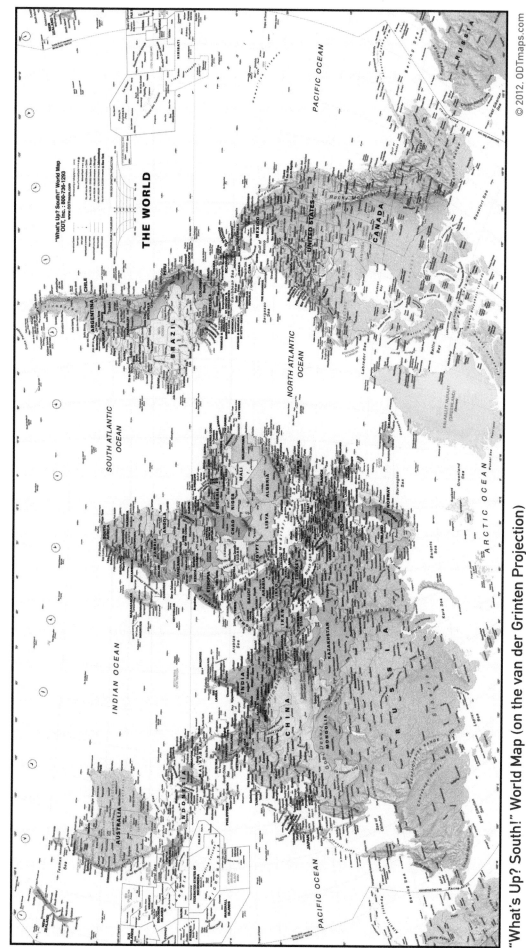

"What's Up? South!" World Map (on the van der Grinten Projection)

THE WORLD

"What's Up? South!" World Map
ODT, Inc. : 800-736-1293
www.ODTmaps.com

15

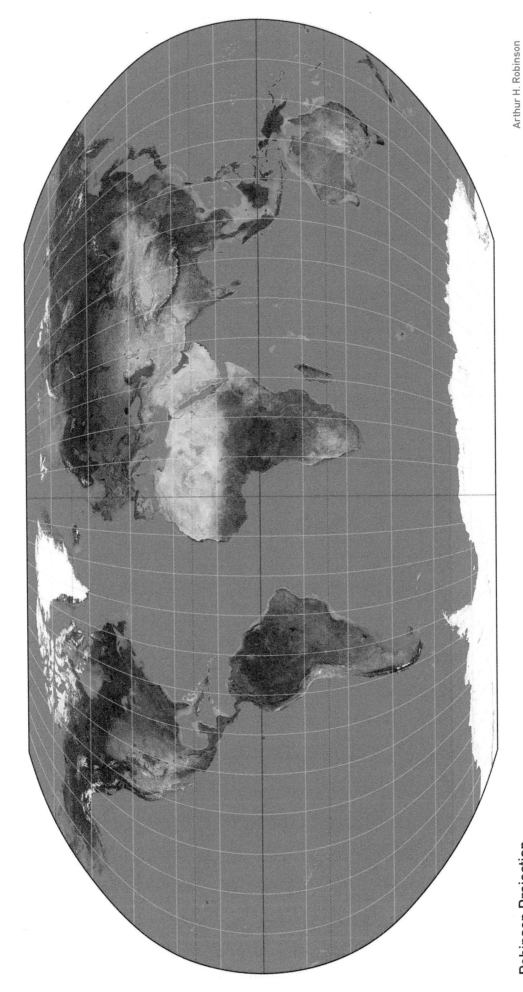

Robinson Projection

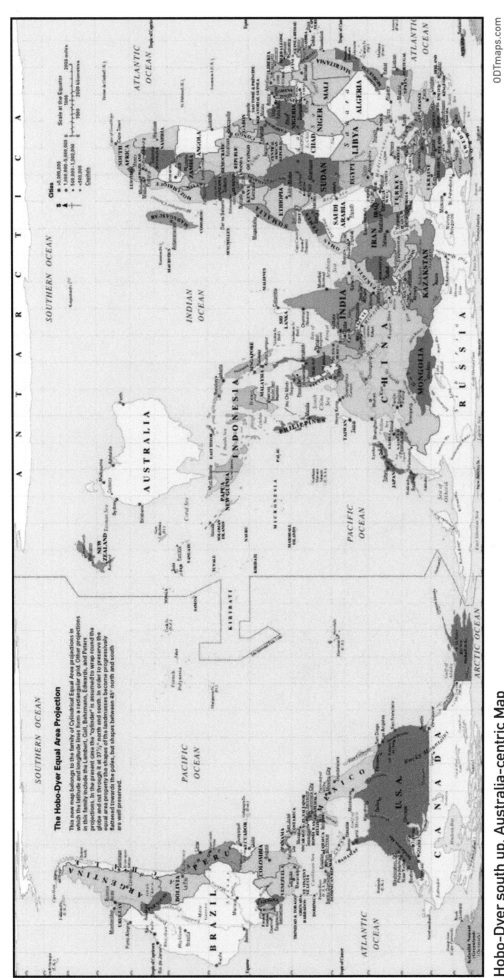

The Hobo-Dyer Equal Area Projection

This new map belongs to the family of Cylindrical Equal Area projections in which the latitude and longitude lines form a rectangular grid. Other projections in this family include the Lambert, Gall, Behrmann, Edwards, and Peters projections. In the present case, the "cylinder" is assumed to wrap round the globe and cut through it at 37½° north and south. In order to preserve the equal area property the shape of the landmasses become progressively flattened towards the poles, but shapes between 45° north and south are well preserved.

Hobo-Dyer south up, Australia-centric Map

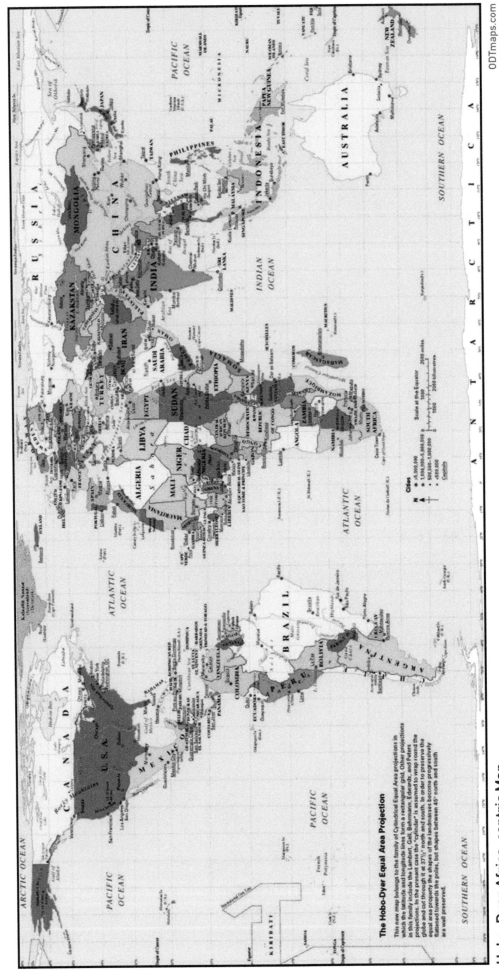

The Hobo-Dyer Equal Area Projection

This new map belongs to the family of Cylindrical Equal Area projections in which the latitude and longitude lines form a rectangular grid. Other projections in this family include the Lambert, Gall, Behrmann, Edwards, and Peters projections. In the present case the "cylinder" is assumed to wrap round the globe and cut through it at 37½° north and south. In order to preserve the equal area property the shapes of the landmasses become progressively flattened towards the poles, but shapes between 45° north and south are well preserved.

Hobo-Dyer Africa-centric Map

ODTmaps.com

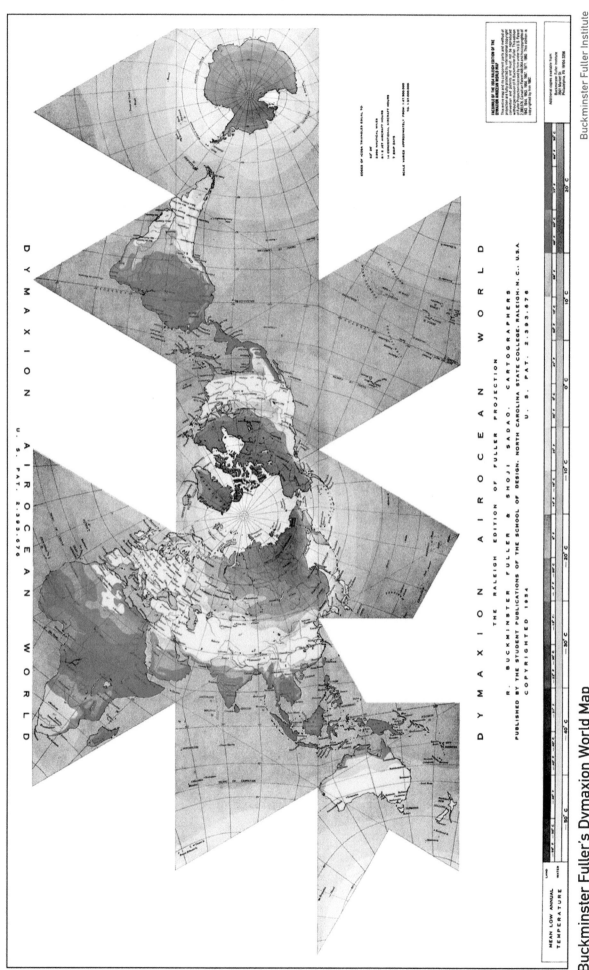

Buckminster Fuller's Dymaxion World Map

Buckminster Fuller Institute

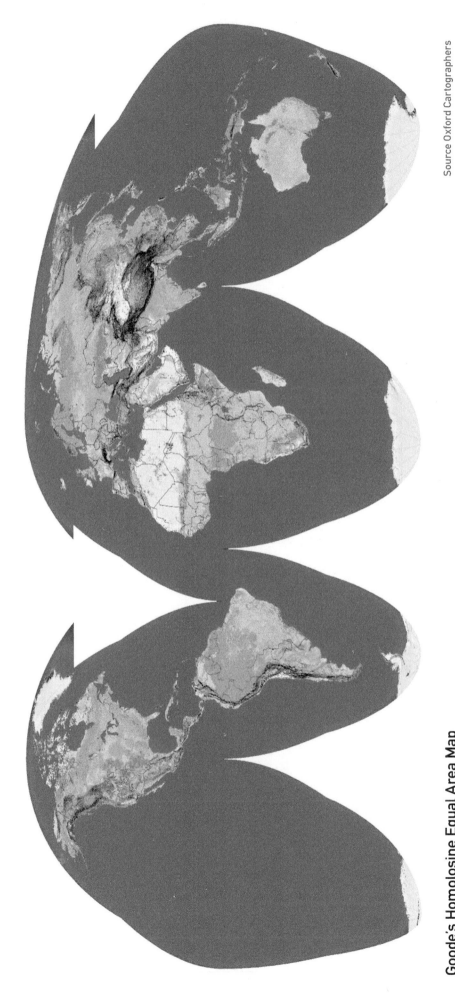

Source Oxford Cartographers

Goode's Homolosine Equal Area Map

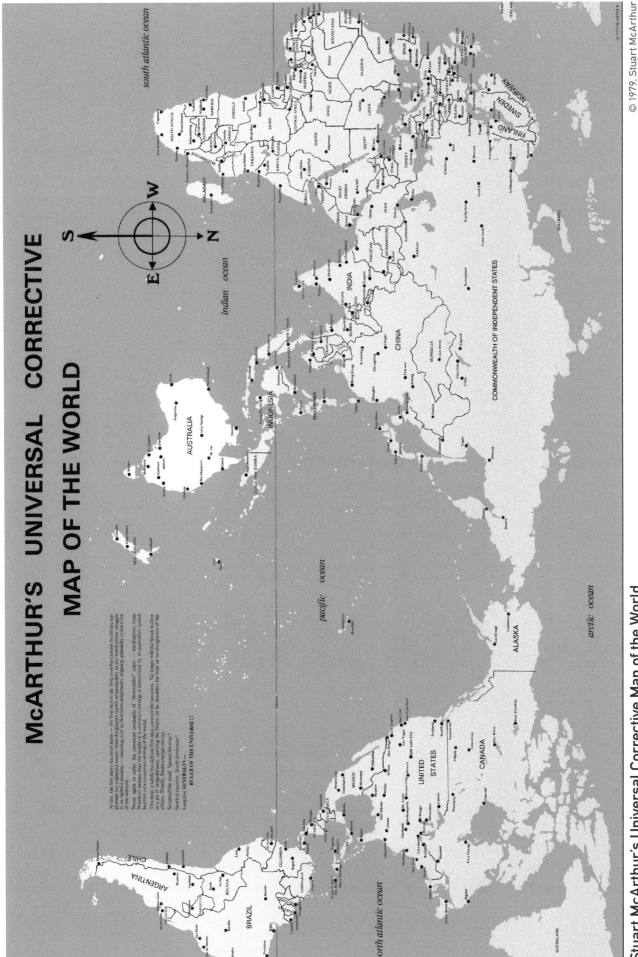

McARTHUR'S UNIVERSAL CORRECTIVE
MAP OF THE WORLD

At last, the first move has been made — the first step in the long overdue crusade to elevate our glorious but neglected nation from its gloomy depths of anonymity in the world power struggle to its rightful position — towering over its Northern neighbours, reigning splendidly at the helm of the universe.

Never again to suffer the perpetual onslaught of "downunder" jokes — implications from Northern nations that the height of a country's prestige is determined by its equivalent spatial location on a conventional map of the world.

This map, a subtle but definite first step, corrects the situation. No longer will the South wallow in a pit of insignificance, carrying the North on its shoulders for little or no recognition of her efforts. Finally, South emerges on top.

So spread the word. Spread the map!

South is superior. South dominates!

Long live AUSTRALIA —
— RULER OF THE UNIVERSE!!

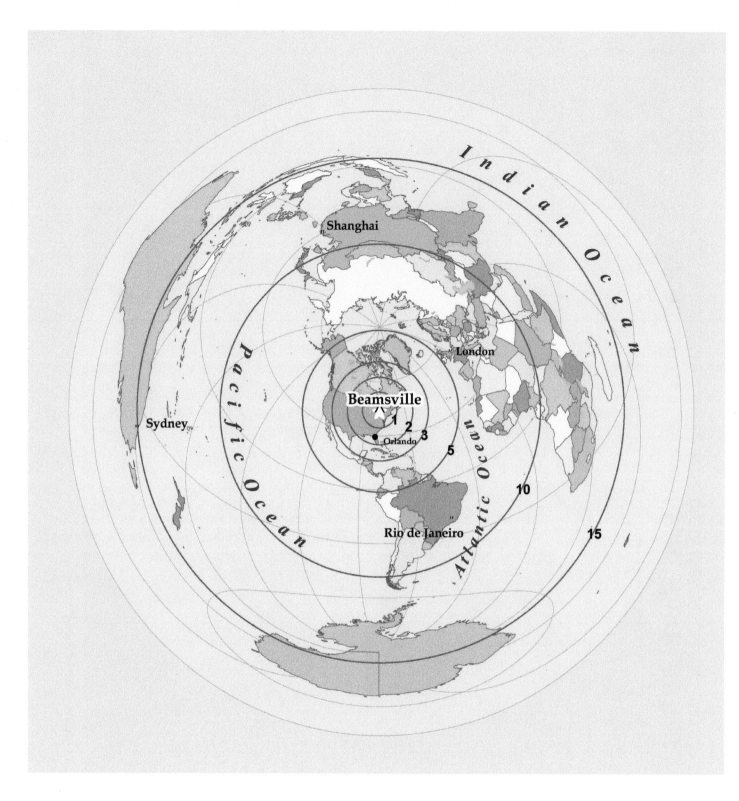

Azimuthal Equidistant Map – centered on Ward Kaiser's home town of Beamsville, ON